BUSINESS/SCIENCE/TECHNOLOGY DIVISION
CHICAGO PUBLIC LIBRARY
400 SOUTH STATE STREET
CHICAGO, IL 60605

R001 5 0 7789

CI
HAROLI

D0919505

R001

BUSINESS & INDUSTRY DIVISION

FORM 125 M

cop. 2

The Chicago Public Library

Received_____Aug 29, 1973_____

THE CHICAGO
PUBLIC LIBRARY

GROWING AGAINST OURSELVES:
THE ENERGY-ENVIRONMENT TANGLE

Publications of the John F. Kennedy Institute,
Center for International Studies, Tilburg, the Netherlands
Nr. 6.

GROWING AGAINST OURSELVES:
THE ENERGY-ENVIRONMENT TANGLE

Problems, Policies and Approaches

Edited by:
S. L. Kwee
J. S. R. Mullender

With contributions from:
Francis A. Beer
J. Clarence Davies
Future Shape of Technology Foundation
(J. H. Bakker, J. A. G. Davids, H. van Duuren, A. J. Elshout,
J. A. Goedkoop, J. C. ten Houten, P. E. Joosting, K. J. Keller,
J. L. Koolen, M. Muysken, J. J. Went)
Jacek Janczak
Peter-Jörg Jansen
R. J. H. Kruisinga
S. L. Kwee
Robert M. Lawrence
J. M. Martin
Pawel Jan Nowacki
Dean Schooler
Craig Sinclair

and a foreword by:
Frans A. M. Alting von Geusau, Director John F. Kennedy Institute

English language consultant:
Mrs. Nanette Gilmour

A. W. SIJTHOFF-LEIDEN
1972

Lexington Books
D.C. Heath and Company
Lexington, Massachusetts
Toronto   London

Cop. 2

Published in the United States of America by Lexington Books, D. C. Heath and Company, Lexington, Massachusetts.

ISBN 90 286 0152 X (Sijthoff)
ISBN 0-669-84715-1 (Lexington Books)

Library of Congress Catalog Card Number: 72-79804
   A. W. Sijthoff International Publishing Company, N.V. 1972.

No part of this book may be reproduced in any form by print, photoprint, microfilm or any other means without written permission from the publisher.

Printed in the Netherlands.

THE CHICAGO PUBLIC LIBRAR.
AUG 29 '73 B

CHICAGO PUBLIC LIBRARY

FOREWORD

The increasing impact of scientific and technological developments on society has recently become a major concern for man, social groups, policy-makers and the scientific community. Among the problems raised by these developments, the political problem of making choices, formulating criteria, establishing rules, allocating resources and mastering effects can be singled out as among the more crucial ones. The need for more responsible and more imaginative national and international policies has been widely felt but inadequately met sofar. The same could be observed with respect to university research in this field. Universities, like governments and international organizations, are vertically organized if not compartmentalized, with insufficient flexibility for approaching major contemporary problems on a multi-disciplinary or inter-agency basis.

In view of this situation, the John F. Kennedy Institute in 1969 decided to embark upon a new multi-disciplinary program in Science, Technology and Public Policy with the aim of contributing to a better understanding of the interaction between developments in the field of science and technology on the one hand and decision-making in the organs of public policy on the other. The program was intended to improve general understanding of the problems of interaction and to contribute to the elucidation of specific and crucial problems faced by experts and policy-makers. The program was organized in such a way that policy-oriented research and regular discussion between those concerned would complement each other in promoting better understanding of the problems involved. One of the specific and crucial problems selected in the initial stage of the program, was the rapidly increasing need for energy and the effects energy production would have for the environment. From June 9 to 12, 1971 the Institute in cooperation with the Future Shape of Technology Foundation in the Hague therefore organized an international colloquium on Electric Energy Needs and Environmental Problems. It was meant to confront energy-producers, experts and politicians with a problem they thus far did not tackle jointly.

The present volume reproduces most of the papers prepared for the colloquium. As such it does not pretend to offer more than a first descriptive and comparative analysis of problems, policies and future strategies. It has to be followed by more focused and more in-depth analyses of at least some of the problems raised.

Part I of the book opens with a revised version of the opening address to the colloquium delivered by the then Netherlands' Health Secretary. It is followed (in Chapter II) by a discussion of the major options the Institute presented for discussion to the participants. Chapter III identifies the international dimensions of the problem of energy and environment.

Part II gives a survey of technical problems and energy policies conducted in some important countries. It opens with basic technical information in Chapter IV. It continues by describing the policies in the United States, the United Kingdom, France and Poland.

Part III, finally, presents some material on future strategies and approaches: Chapter IX on possible American policies, and Chapter X on the need for a systematic analysis of energy-environment planning.

In Chapter XI, the editor comments on a number of future strategies and approaches discussed during the colloquium.

To assist the reader, we have added two recent documents in the annexes: the message of President Nixon to the U.S. Congress outlining a program to ensure adequate supply of clean energy for the years ahead and a document presented to the colloquium by the Commission of the European Community.

The Institute is indebted to the Netherlands' Ministry of Science and Education and the United States Educational Foundation in the Netherlands for their financial support towards the colloquium. It is indebted also to Sijthoff International Publishing Company for publishing and to Heath and Company for co-publishing this volume.

The Institute is particularly grateful to Prof. S. L. Kwee and Ir. J. S. R. Mullender for editing this volume and to Mrs. Nanette Gilmour for correcting the manuscripts and for preparing the index. It is equally grateful to the authors for having contributed to this volume. Special words of gratitude should be addressed to Prof. Dean Schooler, who has been the "mastermind" behind this effort and to the Future Shape of Technology Foundation, which has given permission to reproduce an abridged version of their publication no. 7 in this volume.*

---

* This Foundation (Stichting Toekomstbeeld der Techniek) was established in 1968 by the Royal Institution of Engineers in the Netherlands. Its purpose is to study future technological developments and, where possible, to illuminate social consequences. In order to contribute to an integration of modern technology in future society monographs are being published, dealing with the various subjects of study. Publications and information: Stichting Toekomstbeeld der Techniek, Prinsessegracht 23, The Hague, The Netherlands, phone 070-646800.

Without the invaluable and unwearied assistance of Miss Hinkenkemper, secretary to the Institute and Miss Vugs, assistant to the secretary, this volume could not have been published.

August 1971
Frans A. M. Alting von Geusau,
Director of the Institute

TABLE OF CONTENTS

x

ABOUT THE AUTHORS

S. L. Kwee (editor and Chapter XI) is Professor of Philosophy at the Technological University of Eindhoven, the Netherlands.

J. S. R. Mullender (editor) is Adviser to the Secretariate of the Science Policy Council of the Netherlands, The Hague.

J. H. Bakker (Chapter IV) is Director Joint Institutions and Laboratories of the Electric Utilities in the Netherlands, Arnhem.

Francis A. Beer (Chapter III) is an associate Professor at the University of Texas, Austin, U.S.A. From February to July 1971 he was a Fulbright-Hays Advanced Research Scholar at the John F. Kennedy Institute in the Netherlands.

J. A. G. Davids (Chapter IV) is Biologist in the Health Protection Department, Reactor Centrum Nederland, Petten, The Netherlands. Publications on radioactive waste disposal and biological accumulation of radio-isotopes.

J. Clarence Davies, 3d (Chapter V) is Senior Staff member, President's Council on Environmental Quality, Washington, D.C., and the author of "The Politics of Pollution" (New York: Pegasus, 1970).

H. van Duuren (Chapter IV) is Chemical engineer, Joint Laboratories of the Electric Utilities in the Netherlands, Arnhem. He is co-author of several articles on the environmental effect of fossil fuel combustion.

A. J. Elshout (Chapter IV) Research and development officer, Joint Laboratories of the Electric Utilities in the Netherlands, Arnhem.

J. A. Goedkoop (Chapter IV) is Managing Director for Research, Reactor Centrum Nederland, Petten, and Extraordinary Professor of Physics at the University of Leiden, the Netherlands.

J. G. ten Houten (Chapter IV) is Director of the Institute for Phytopatho-logical Research and Extraordinary Professor on Air Pollution Control; Agricultural University, Wageningen, the Netherlands.

Jacek Janczak (Chapter VIIIA) is Director of the Air Pollution Control Office of Central Water Resources Administration in Poland.

Peter-Jörg Jansen (Chapter X) is heading the group "Systemtechnik" at the "Institut für Angewandte Reaktorphysik", Kernforschungszentrum, Karls-ruhe (K.F.K.). He has contributed to the Symposium on Environmental Aspects of Nuclear Power Stations, New York, August 1970, I.A.E.A.-S.M.-146/57, Vienna 1971. January 1972 (K.F.K. 1511), contribution to an analysis of the cooling capacity of the river Rhine (Nuclear Stations).

P. E. Joosting (Chapter IV) is Medical Research Officer and Adviser TNO-Research Institute of Public Health Engineering, Delft, the Netherlands. Publications on air quality criteria and standards; e.g. American conceptions as compared to Russian notions (1966), etc.

K. J. Keller (Chapter IV) is Senior Physicist Joint Laboratories of the Electric Utilities in the Netherlands, Arnhem.

J. L. Koolen (Chapter IV) is Head of the Department of Surface Water, Government Institute of Sewage Purification and Industrial Waste Treat-ment, Voorburg, the Netherlands.

R. J. H. Kruisinga (Chapter I) was at the time of the Colloquium Secretary of State of Social Affairs and Public Health in the Netherlands, The Hague.

Robert M. Lawrence (Chapter IX) was at the time of the Colloquium Asso-ciate Professor at the University of Arizona, Department of Government, Tucson, Arizona. He is now Professor at Colorado State University, De-partment of Political Science. In cooperation with Gaylon Caldwell he published American Government Today (New York: Norton, 1969).

J. M. Martin (Chapter VII) is Research Officer at Social Sciences University of Grenoble, Institute of Economic Research and Planification, Institute of Economic and Legal Studies on Energy, St. Martin-d'Hères, France. Pu-blications concerning industrial developments in South America, 1967; "Energy Intensive Industry", study-group on agro-industrial complexes, I.A.E.A., Vienna, 1971 (19p).

M. Muysken (Chapter IV) is Mechanical Engineer co-ordinating the light water reactor development, Reactor Centrum Nederland, Petten, the Netherlands.

Pawel Jan Nowacki (Chapter VIIIB) is director of the Scientific Center of the Polish Academy of Sciences, Paris.

T. C. Sinclair (Chapter VI) is Senior Research Fellow, Science Policy Research Unit, University of Sussex, Falmer, Brighton, Great Britain, and co-editor of Research Policy, quarterly of North-Holland Publishing Company, Amsterdam. Publications on R & D-management, Anti-Pollution Programming, etc.

Dean Schooler Jr. (Chapter II) is an Assistant Professor of Government and Associate Director for Research Development of the Institute of Government Research at the University of Arizona (Tucson, Arizona, U.S.A.) During 1970–1971 he was a Fulbright-Hays Advanced Research Scholar at the John F. Kennedy Institute in the Netherlands. He has written on environmental policy and authored a book, Science, Scientists and Public Policy, (New York: The Free Press, 1971).

J. J. Went (Chapter IV) is Director Joint Institutions and Laboratories of the Electric Utilities in the Netherlands, Arnhem and Extraordinary Professor of Reactor physics at the Technological University of Eindhoven, the Netherlands.

ABBREVIATIONS

ACORD	Advisory Council on Research and Development (U.K.)
AEC	Atomic Energy Commission (U.S.A.)
AEA	Atomic Energy Authority (U.K.)
BOD	Biochemical Oxygen Demand
BWR	Boiling Water Reactor
CBA	Cost-benefit Analysis
CEGB	Central Electricity and Gas Board (U.K.)
CERN	European Organisation for Nuclear Research
CMEA	Council for Mutual Economic Assistance (E. Europe)
DATAR	Directorate for Site Planning (Fr.)
DDR	German Democratic Republic
DIME	Mechanical and Electrical Industries Direction
DGRST	General Directorate for Scientific and Technological Research (Fr.)
ECSC	European Coal and Steel Community
EDF	Electricité de France (Nat. Public Corporation)
EEC	European Economic Community
ENEA	European Nuclear Energy Agency
EPA	Environmental Protection Agency (U.S.A.)
EURATOM	European Atomic Energy Community
FAO	Food and Agriculture Organisation
FPC	Federal Power Commission (U.S.A.)
FRG	Federal Republic of Germany
FY	Fiscal Year
GNP	Gross National Product
GPO	Governmental Printing Office
HMSO	Her Majesty's Stationery Office
IAEA	International Atomic Energy Agency
ICC	Interstate Commerce Commission (U.S.A.)
ICRP	International Commission on Radiological Protection
IEA	International Environmental Agency

KEMA	Joint Laboratories of the Electric Utilities in the Netherlands
KNMI	Royal Netherlands' Meteorological Institute
LMFBR	Liquid Metal Fast Breeder Reactor
LNG	Liquid Natural Gas
MHD	Magneto Hydro Dynamic (Generators)
MIT	Massachusetts Institute of Technology
NCB	National Coal Board (U.K.)
NCR	National Council of Research (Fr.)
OECD	Organization for Economic Cooperation and Development
OST	Office of Science and Technology (U.S.A.)
PAN	Peroxy-acetylnitrate
PWR	Pressurized Water Reactor
RH	Relative Humidity
RFE	French Energy Review
R&D	Research and Development
SALT	Strategic Arms Limitation Talks
SNCF	National Society of Rail-roads (Fr.)
TCE	Tons of Coal Equivalent
TNO	Research Institute of Science and Technology (NL)
TVA	Tennessee Valley Authority
UK	United Kingdom
UNCTAD	United Nations Conference for Trade and Development
UNESCO	United Nations Educational Scientific and Cultural Organisation
UNIPEDE	Union Internationale des Producteurs et Distributeurs d'Energie Electrique
USA	United States of America
USSR	Union of Socialist Soviet Republics
kWh	10^3 Wh
MWh	10^6 Wh
GWh	10^9 Wh
TWh	10^{12} Wh

Part I

ELECTRIC ENERGY AND ENVIRONMENT: IDENTIFYING THE PROBLEMS

Chapter I

INTRODUCTION

by *R. J. H. Kruisinga*

The current dynamic development of economics and technology relegate projections of future electrical energy needs virtually to the realm of crystal ball gazing. Anyone concerned clearly faces new and important challenges. Highly specialized electrical energy technologists are now being asked to widen their outlook with a comprehensive view of the environmental implications of their work. Conservationists who find more question marks than desirable solutions in new technical developments are pressed in return to develop more realistic evaluations of energy needs before passing final judgment on future projects. Only by such a broadening of attitudes will acceptable solutions be found. Consequently, an important question arises: "How is a fruitful confrontation between contending views of energy needs to be organized?" Talking together and, above all, thinking together is essential. That in itself makes this colloquium useful. For only a reasoned and comprehensive view of the future is of any use.

As we press further in our examination of the subject, the lines of divergence and increasing specialization become even more apparent. With respect to environmental quality, there are essential and obvious differences between the use of coal, oil, natural gas, and nuclear power as primary sources of energy. Few, if any, have such a grasp of objective criteria as to enable them to weigh the advantages and disadvantages of the various methods and thus to arrive at responsible conclusions. Aren't we here, then, touching upon a second important problem in modern society: How can we arrive at comprehensive and responsible conclusions?

For example, how must the danger of air pollution by sulphur dioxide (SO_2) be compared with the danger of the nitrogen oxides? How should we compare, to make matters even more difficult, the radioactive inert gases that a nuclear power station discharges into the atmosphere with SO_2 pollution? A problem of still a different nature is that of heat emissions by power stations, particularly by nuclear energy plants. What effects can this —thermal pollution—be expected to have on the ecosystems of adjacent waters? Should an arbitrary limit of 30° C be set as the allowable discharge temperature? If so, how would one justify 30° C?

In many countries where the once-through cooling system can be effectively used in only a few places, thermal pollution is a serious problem. Other methods of cooling discharge water from nuclear plants have been developed, but they have inherent problems as well. A system of cooling towers has little to recommend it in aesthetic terms. Moreover, use of the cooling water for other purposes such as heating has limited utility because of the relatively low temperatures attained. Beyond this thermal pollution, however, nuclear power stations are at an advantage. The danger of radioactive harm through regular or accidental emissions has been minimized sufficiently so as to not present an irresponsible risk. But one of the most serious difficulties remaining is that of reprocessing irradiated fissionable materials.

In order to meet the long-term requirements of energy needs considerable discussion and many compromises among the opinions held by concerned scientists of various disciplines, is essential. The public must also be included in these discussions. Frank information and strong arguments concerning safety procedures, environmental damage, costs, outputs and so forth must be realistically presented if public apprehension and opposition is to be allayed. Indeed, the permit procedures of many countries already include public hearings as an essential component.

A total conversion to nuclear power generation cannot as yet be regarded as a real alternative, but these three points appear quite certain: First, the electrical energy needs of society are rising very rapidly. Second, factors limiting the production of electrical energy will not be capital investment or size of labor force, but accompanying costs to the environment. Finally, counter measures for current environmental threats are vital for the survival of society itself. Solution of these problems must be considered of the utmost importance.

The specific problems which serve as a focus for this volume almost inescapably evoke fundamental questions for society at large:

1. Fusion reactors and fast breeder reactors are possible technological alternatives to the use of coal, fuel oil, and conventional nuclear power in meeting rapidly rising demands for electricity with minimal harm to the environment. Is fusion (relying on deuterium abundantly available in sea water) a realistic and viable alternative for power production in the near future? What methods of power production are being used or contemplated by the various nations?

2. Pollution apparently knows no political boundaries, appearing as a by-product of capitalist, socialist, and state-owned electrical power production facilities. Does any political or economic system offer a better method for reconciling the demand for energy and the demand for a clean and livable environment?

3. What are the actual and desired roles of power-producing industries, whether state or privately owned, when an energy or environmental problem arises? Do powerful energy-oriented interest groups threaten attempts to guard the environment?

4. Governments can choose from a variety of instruments in balancing energy needs and environmental preservation—subsidies, required abatement procedures, controls on site location of production facilities, price controls on the forms of energy and uses, safety regulations, support for research and development, selective taxes, and so on. Which methods or combinations of methods are the most useful for different governments and for varying industrial or environmental situations?

5. Can the energy consumption curve continue indefinitely upward? Where must it stop, and how can demands be limited?

Discussion of these points is of major importance and should be a help in guiding governments attempting to set policies in this area.

Although passage of specific legislation and appropriation of funds needed in the fight for environmental protection are within the domains of individual countries, there can be no doubt that environmental problems are of international dimension and importance. Polluted air never stops at national frontiers, however well fortified; foul river water flows untrammeled past the most scrupulous customs inspectors. However, the impediments to dealing with these problems on an international basis are numerous.

Systematic international cooperation, both to meet soaring energy demands and to reduce current dangers to the environment will be extraordinarily difficult to achieve. It would, however, make possible a substantial concentration of available financial resources for these tasks and an acceleration in the exchange of scientific findings. The latter is especially important because of the current acute shortage of scientific knowledge, equipment, and personnel in this field. Naturally, a serious effort of this sort will threaten aspects of the short-term economic positions of some countries unless great care is taken to assure the equalized impacts of new technology deployments. But environmental dangers forecast by expansion of current electrical production technologies are so dire that no single national economy will be able to cope with them unassisted. Uncoordinated development of technology in this field by individual nations is a luxury that can no longer be permitted. International cooperation in the development and widespread application of clean energy production technologies is now an overwhelming and immediate necessity.

Chapter II

THE POLITICS OF ENVIRONMENT AND NATIONAL
DEVELOPMENT: BASIC OPTIONS AND POLITICAL
RESPONSES TO ENVIRONMENTAL PROBLEMS

by *Dean Schooler, Jr.*

The rising tide of political debate surrounding oil spills, detergents, automotive emissions, industrial wastes, municipal sewage and a myriad of other
pollutants, including substances associated with the production of electrical
energy, brings nations to a new stage in the relationship of environment to
national development.[1] These movements from one stage to another have
significant implications for the options, political situations, and strategies
faced by environmental policy-makers. What are these emplications? What
governmental solutions are associated with particular periods? How are
policy-makers and societies' basic options limited? What means and techniques are available to government for balancing economic growth (resource
use, energy needs, rising consumer demands) with environmental quality
(land use, air and water pollution, waste disposal)?

Political systems may go through three major eras in the relationship of
development to environment. The three eras can be distinguished in terms

1. Several authors have contributed basic orientations to this effort. Conceptually, there
is a debt to Allen Schick, "The Cybernetic State", *Trans-action* (February, 1970), pp. 15–
26; Theodore J. Lowi, "American Business, Public Policy, Case Studies and Political
Theory," *World Politics*. XVI: 4 (July, 1964), 677–715; Thomas Vitullo-Martin, "Pollution
Control Laws: The Politics of Radical Change," in Leslie Roos, Jr., *The Politics of Eco-
suicide* (New York: Holt, Rinehart and Winston, 1971), pp. 346–367; and correspondance
with Robert Gilmour, cf., "Private Interests and Public Lands," *Current History*, 59; 347
(July, 1970), 36–42, 52. Substantively, three basic sources are Energy Policy Staff, Office of
Science and Technology, *Electric Power and the Environment* (Washington: Government
Printing Office, August 1970); Environmental Policy Division, Legislative Reference
Service for the Joint Economic Committee, U.S. Congress, *The Economy, Energy and the
Environment*, 91st. Congress, 2nd. Session (Washington: Government Printing Office,
September 1, 1970); and J. Clarence Davies, *The Politics of Pollution* (New York: Pegasus,
1970). The John F. Kennedy Institute International Colloquium on "Electrical Energy
Needs and Environmental Problems" (Eindhoven, The Netherlands, June 1971) and the
contributions of other authors in this volume have been even more immediately informative. This effort strives toward a theory of environmental policy-making set in a more general socio-economic context. Such a theory would provide an overall viewpoint from which
to consider the current debate over electrical energy needs and environmental problems
(brownouts, clean air and resource consumption).

of their different political arenas (environment and development policy-making), political situations, and economic conditions. As stages or eras, they can be characterized as:

The Political Economy of Expansion and Resource Development
The Political Economy of Environmental Confrontation
The Political Economy of Ecology

Most developed nations are moving from the "resource development" phase into a period of "environmental confrontation," but they have a long way to go before entering an era of "ecology." The developing nations remain locked in a justifiable preoccupation with "resource development," and pollution abatement comes behind economic growth and the abolition of poverty as an objective. Moving from the stage of "resource development" to "environmental confrontation" or on to "ecology" depends on (1) citizens and policy-makers' *perception of the severity of environmental problems* and (2) *achievement of a level of affluence and surplus* which will permit paying for a clean environment and devoting time to environmental action.[2]

Presumably, most nations or systems begin with a task of "resource development" and a political economy adapted to that objective. Government's role involves the expansion of the economy, promotion of resource exploitation and use, and encouragement of development through public projects and the private sector. Gradually, attainment of a sufficient level of affluence enables concern for the "quality" of life and environmental pollution. Thus, new values and objectives come into play in a political-economic situation of "environmental confrontation". Economic growth, exploitation of "unlimited" resources, and the free use of unpolluted air and water are challenged by environmentalists, and conflict results. Should these "confrontations" produce programs which successfully abate or maintain existing levels of pollution, then perhaps it will be unnecessary to move into the final third phase—a political economy of ecology. But if they are not sufficient, and either pollution increases or citizens' tolerance for existing levels

---

2. Developing countries and "poverty" areas such as urban ghettos and regions built around declining industries are more preoccupied with economic development than pollution abatement. Jobs and material goods take precedence over clean air and water, and energy needs dwarf the environmental costs of meeting them. Indeed, environmental controls may mean fewer new factories in a poorer area, continuing unemployment, and no increased production. Furthermore, insofar as environmental controls raise the price of products, the poor may bear a disproportionate burden since (1) they consume more goods than non-polluting services and (2) goods with highly-polluting production processes may comprise a larger percentage of the lower income budget. So environmental controls may be "regressive" in impact and, for many in the world, may be harmful. Such regions and areas need a continuing "political economy of resource development" until they reach a level of affluence which will enable them to worry about pollution.

decreases, then we might well move into a more comprehensive phase in which cooperation replaces conflict in the face of a more serious environmental challenge.[3]

These comprehensive factors limit societies' options. Social systems preoccupied with economic development are less likely to develop pollution control technologies and even more unlikely to spend public funds developing expensive, though less polluting, means of producing goods and services. Systems which do not consider the problem serious enough or which do not have the necessary surplus resources are unlikely to embark on *radical transformations to new methods of producing electrical power* (such as fast breeders, fusion facilities, solar plants). These innovations will come, but only with time and sufficient funds. The environmental "crisis" is not severe enough and the resources are not great enough to justify a "crash" program. Instead, government will continue incrementally to regulate polluters and abate pollution within the confines of existing and near-future technology. Nor are most societies at the stage of seriously limiting economic growth, energy consumption, or the production of goods and services. Perceptions of the problem, political reality, and the constraint of limited surplus resources make an incremental response inevitable and radical solutions unlikely, but this could change.

Simply stated, we are concerned with the limitations which the nature of the problem and resources of the economic system place on the options, strategies, effective criteria, governmental programs, and techniques involved in environmental policy-making.

The Political Economy of Expansion and Resource Development

Political systems striving to move from economies of scarcity to economies of affluence have a distinct relationship to environment. Their objectives are growth and development-oriented, and are directed toward:

1. greater variety, diversity and available life styles in the society
2. increased rates of economic growth and rising living standards
3. larger tax bases and increased government revenue
4. greater national security
5. protection for unique geological or archeological areas such as national parks and monuments

3. This framework is influenced heavily by Allen Schick, "The Cybernetic State," *Transaction* (February, 1970), pp. 15–26. These eras or periods are not mutually exclusive. Attitudes, governmental programs, techniques of environmental control, and strategies carry over from one period to the next. Still, the new importance of new attitudes, programs, techniques and strategies combined with the lesser role of older aspects gives a distinct, different quality to the new era or "political economy". The new forms are dominant.

6. development and "use" of all other "non-unique" lands and resources through stimulation of individual initiative and the market mechanism

The society's preoccupation with "development" and "expansion" dwarfs the concern of a minority for a feared scarcity of non-renewable resources. Indeed, their protests do not disturb the society's basic assumption of an inexhaustible supply of basic resources. "Conservation" seems only effective when tied to a concern for national security (petroleum reserves, oil import quotas, and so on).

Development and expansion are achieved through the initiative of individual sectors of the society, competing with one another and pursuing their own maximum self interest. Various sectors—mining, grazing, recreation, public and private power, agriculture, industry—are involved in a race for the fastest rate of growth. Government agencies, each tied to a particular sector, either stimulate or promote the efforts of their "clientele" or engage in "development" themselves (cf. public power, the Bureau of Reclamation and Tennessee Valley Authority).[4]

Environmentalists, despite their lack of effectiveness, are themselves fragmented into sectors which compete with one another. Narrow "mission" orientations insulate conservation-minded government bureaus and different interests divide conservationist groups.[5] No "ecological" perspective unites them, and instead they compete to develop for "use" whatever areas they need for their purposes. Objectives of "multiple use" may merely rationalize such real divisions among interests in the society—a way of agreeing to disagree, when there seems to be room for anyone to pursue his own desires.

The "politics" of expansion and resource development constitute a *positive non-zero sum game* for groups in the society.[6] Everyone pursues his own objectives at no or minimal perceived cost to others, and the whole society gains from the self-interested initiatives of individuals and groups. Politically, the situation could be described as a pork barrel operation facilitated by logrolling and backslapping. Inter-group conflict is minimized and

4. The purpose of promoting private development of electrical power or government itself building production facilities would be an unlimited supply of cheap electricity. This supply would be essential to continued and increased economic growth and development.
5. The divisions and narrow mission orientations among government agencies are discussed in Ashley Schiff, "Innovation and Administrative Decision-Making: The Conservation of Land Resources," *Administrative Science Quarterly*, II:1 (June, 1966), 1–32.
6. Such a situation existed in the United States from its origins, but particularly from 1900 through the late 1960s. Policy-making regarding environment and national development has only recently begun to assume the character of a political economy of "environmental confrontation."

government is unlikely to coerce individuals or groups.[7]

Policies are formed and adapted to the needs of individual sectors. The locus of policy-making resides within legislative committees and executive agencies, and as a result, policies are likely to be *ad hoc*, segmented, and not based on long range, system-wide planning. Such a decentralized system, with each component bent on promoting and expanding its own area, cannot develop a coherent, unified policy.

The bulk of political power is held by (1) single-purpose, self-interested, tightly-organized economic groups and (2) geographical and political sub-divisions whose constituencies demand projects and government funds as payoffs for their votes. These specific interests, tied variously to economic self-interest or government's need for tax revenues, have a political advantage over less powerful, unorganized, and diffuse groups. Such groups—consumers, "breathers," or adjacent property owners—probably do not worry about "second-order consequences" for themselves. But even if they did not place more value on economic growth, their power would be limited.[8] Regardless, the general public is not at all involved in environmental or developmental activities in an era of expansion and resource development.

The activities of government in an era of expansion and resource development are basically "distributive" in character.[9] Policies may involve (1) the promotion and assistance of private enterprise; (2) actual government development of selected resources and economic areas; and (3) the turning-over of public goods (land, air and water) to private entrepreneurs or managers of specific enterprises for their use free-of-charge. While government may have begun with a hands-off, laissez-faire policy of "limited government," it gradually assumed a more active role in the development process.[10] That role was, however, in both cases essentially "distributive" in terms of values, benefits and rights associated with economic development and growth.

Government's "distributive" actions may benefit both developmentalist and environmentalist groups. Land leasing arrangements, homesteading

7. Theodore Lowi, "Decision-Making and Policy-Making: Toward an Antidote for Technocracy," *Public Administration Review*, 30:3 (May-June, 1970), 314–325.

8. Mancur Olson, Jr., *The Logic of Collective Action: Public Goods and the Theory of Groups* (New York: Schocken Books, 1968).

9. Theodore Lowi, "American Business, Public Policy, Case Studies, and Political Theory," *World Politics*, XVI:4 (July, 1964), 677–715. Lowi's initial introduction of the concepts of "distributive," "redistributive," and "regulative" arenas forms a base for much of this argument.

10. Allen Schick, "The Cybernetic State," *Trans-action* (February, 1970), pp. 15–26. Schick's "political state" assumes a limited, laissez-faire notion of government and his "bureaucratic state" assumes an active role for government as "doer." Both states co-exist with the politics of expansion and resource development.

laws, depletion allowances (oil, various minerals), and investment tax credits benefit expanding industries and economic development. Although much less prevalent and substantial than pro-expansion actions, government actions may also benefit environmentalist groups—even amid an era concerned with economic expansion. Such a "distributive" approach to environmental policy takes the form of *subsidies* for pollution abatement systems (sewage treatment facilities in municipalities) or accelerated depreciation allowances and tax credits (effluent control systems for industries).[11] However, the "distributive" approach may be a "failure" in abating or limiting the growth of pollution, since it depends upon the polluter's initiative and provides no incentives for further effluent reduction or increases in the price of goods produced to pay control costs.[12] *Standards*, even if they exist in government's environmental policy arsenal during an era of expansion, are likely to be (1) unenforced, (2) not very demanding, and (3) established in line with either existing levels of pollution or existing abatement technology capability. Such "regulative" or enforcement-oriented instruments would be out-of-step with an era built around expansion stimulated by individual initiative and the government "carrot." Subsidies, tax credits, accelerated depreciation allowances, depletion allowances, and support for private research and development have a more "distributive" aspect and are more in accord with the era's politics.

The "politics" of expansion and resource development prove neither hospitable nor advantageous to environmental groups. Nuisance laws are the major legal base for their efforts, and much of their time would be consumed by litigation in the courts. But both litigation and more direct environmental group activities must be aimed at specific problems (injunctions against one factory, delays in constructing one section of a highway). Concern for the environment remains a "symbolic" phenomenon—an

11. The failure of subsidies as an approach to water pollution control (municipal sewage systems) is argued in a 1969 Report from the United States Comptroller General and Government Accounting Office, "Examination into the Effectiveness of the Construction Grant Program for Abating, Controlling, and Preventing Water Pollution," (Washington: Government Accounting Office, November 3, 1969). Subsidies, political and economic limitations are discussed in A. Myrick Freeman III and Robert H. Haveman, "Water Pollution Control, River Basin Agencies, and Economic Incentives: Some Current Policy Issues," *Public Policy*, XIX:1 (Winter, 1971), 53–74.

12. The major problems with "distributive" (subsidies, tax credits etc.) approaches are that (1) they depend entirely on local initiative; (2) each project is considered alone on its own merits on a first come, first served basis; (3) non-polluters may indirectly pay for someone else's pollution and product prices will not reflect all the costs of production (i.e., pollution); (4) they may lead to unrealistic profits in pollution intensive industries or unwarranted expansion of those industries; and (5) they may encourage technological add-ons when modification of the production process or its inputs might be more effective and economical.

image verbally evoked by many and acted upon by only a few. The general commitment (broad legislation, comprehensive national policies) is directed toward economic expansion and resource development with little regard for environmental concerns.

The Political Economy of Environmental Confrontation

When the society reaches an acceptable level of affluence and its perception of the costs of pollution deems the effects unacceptable, the system moves from a concerted national effort toward economic expansion to a confrontation between the goals of "expansion" and "environmental quality". New objectives rise to the forefront, challenging the assumptions of the expansionist era. These include:

1. concern for the capacity of the environment to absorb or regenerate renewable resources (water, air, soil etc.)
2. attention to the health and aesthetic costs of pollution rather than concern for the conservation of "scarce" non-renewable resources
3. protection from "external costs" for individuals and groups so affected; imposition of these social costs on polluters who previously did not have to calculate such costs in their operations

This shift occurs as a result of a change in the economy—from a situation of material needs and resource abundance to a situation of material affluence and scarcity of clean air and water. Such changes in attitudes toward non-renewable and renewable resources stem from the argument that "...a low-interest, high-construction economy uses up stock mineral resources, and that a high-interest, high consumption economy depends mostly upon renewable biological resources."[13]

The era of environmental confrontation brings with it the need to choose. Groups' demands for environmental quality become as powerful as older groups' desires to develop resources and manufacture goods without regard to environmental considerations. Conflict develops and government variously intervenes on the side of one group or another.

Government's role, however, does have a new aspect—intervening often on behalf of environmental protection and "regulating" the activities of polluters. The positive, non-zero sum game becomes zero sum. There are winners and losers, unless as often happens, compromise gives "polluters" and "environmentalists" each half of their demands. Tangible benefits and losses have at least become notable elements in policy-making, and "environ-

13. Anthony Scott, "Natural Resources: The Economics of Conservation," *Canadian Economic Studies*, No. 3 (Toronto: 1955), p. 96. Scott's point is discussed in Earl F. Murphy, *Governing Nature* (Chicago: Quadrangle Books, 1967), p. 73.

mental quality" is no longer merely a reassuring symbol.[14] Government has stepped in to correct deficiencies in the market system, and to bring "externalities" or social costs into the calculation of product prices. Further, government seeks to create demand for environment-improving and pollution-abating products or technologies.

Policy-making becomes *more centralized* and less isolated in specific sectors (mining, recreation, water pollution, agriculture etc.). This centralization occurs because conflicts begin to emerge between sectors, and these differences can be resolved only on high levels. Such centralized conflict resolution occurs whether the issue divides polluters and environmentalists or divides recreation-oriented environmentalists from wilderness preservationists.

The preferred solution to conflict among groups is compromise achieved through bargaining. This is the most politically attractive solution since "politics" still consists of self-interested groups with intense feelings—and most political leaders would prefer even the losers to win something, later if not immediately. Thus, even when demands for maintaining a clean river conflict with the possibility of meeting increased local electricity requirements and providing jobs for the locally-unemployed with a new power plant, some compromise may be sought if at all possible. No governmental elite is prepared to impose an "all-or-nothing" solution to an environmental confrontation, not when neither the demand for a clean environment nor the demand for economic growth are that powerful or persuasive.

Still, alongside the centralization of conflict resolution, the old sectoral "politics" of the previous era remain. Much remains compartmentalized. Citizens do not see the connection between a burning light bulb and the sulfur oxides and warm water that pour from electrical power plants. Nor do policy makers systematically make comparisons among fuel sources for electricity production, preferring to let each agency promote development of its own area in decentralized fashion (cf. the Department of the Interior with oil and coal, Atomic Energy Commission with nuclear energy, and Federal Power Commission and electric utilities in the United States). Fragmented policy-making does indeed persist alongside the new trends.

The locus of policy-making shifts upward, rising from the agency and legislative committee level to encompass the legislature as a whole and the leaders of the administrative arm of government. The "choices" have become more difficult. Coercion becomes more likely, but is confined to the regulation of individual conduct. Environmental controls are achieved by

14. Murray Edelman, *The Symbolic Uses of Politics* (Urbana: University of Illinois Press, 1964).

regulating individuals' actions rather than changing the conditions in which they behave.[15]

Policy-making remains remedial and long-range policies are rare. Choices are made regarding more immediate and short-term considerations. The thrust of government action is regulative, and takes the form of:

1. emission *standards and criteria* which are enforced through an established, legally-based compliance procedure
2. negotiation with specific polluters to establish standards geared to the economic situation and absorptive capacity of the air or water being polluted in the immediate or regional area
3. establishment of national emission standards for broad categories of polluters (electric power plants, cement plants, sulfuric acid plants, incinerators, automobiles etc.)
4. attempts to centralize the standard-setting and enforcing process
5. direct efforts at regulating corporate polluters and indirect efforts (c.f. automotive emissions) at regulating pollution caused by the average citizen

But regulation does not always develop simultaneously for each form of pollution. J. Clarence Davies argues that:

"...in water pollution the government can offer some kind of service, namely, centralized treatment, but that in air pollution the government is almost exclusively a regulator and coercer."[16]
"...scarcity of public funds is perhaps the major problem in controlling water pollution, whereas in air pollution political power to enforce regulations and the degree of voluntary compliance will be the deciding factors."[17]

Therefore, while solutions to water pollution problems may remain "distributive" (political economy of expansion and resource development), the response to air pollution may be "regulative" (political economy of environmental confrontation).

Government's new activism takes on the character of the "bureaucratic state" described by Allen Schick.[18] National solutions rather than local solutions to problems are increasingly sought. Geographical boundaries become less relevant, and solutions conform more to functions (water basins, air sheds etc.) than traditional territorial responsibilities. Political action seeks to compensate for imperfections in the market, accounting for costs previously unconsidered. Considerations of efficiency and "cost-effectiveness" become factors affecting policies theretofore affected only by constituencies and politically-significant interests. And, as pointed out elsewhere, scientists develop closer relationships with environmental policy-makers who need their advice in successfully "regulating" polluters and

15. Theodore Lowi, "Decision-Making...," pp. 314–325.
16. J. Clarence Davies, *The Politics of Pollution*, pp. 195.
17. *Ibid.*
18. Schick, "The Cybernetic State," pp. 15–26.

their technology in providing a polluter with inexpensive, simple, technological "fixes" for meeting emission standards.[19]

Environmentalists' activities focus on the legislative process and the upper levels of the administrative branch. Regulatory policies are made here, and so litigation in the courts takes a back seat behind legislative and standard-setting processes they must try to influence. The government becomes the prime user of the courts, as a means of obtaining compliance or fines once the administrative remedies of hearings, consultation, and compromise fail. Bur for environmentalists the major regulatory decisions (permits granted to water polluters, site selection and approval for power plants, emission standard setting) are made in an administrative and legislative framework, not in a judicial setting.

The Political Economy of Ecology

The era of environmental confrontation may never evolve into an era of ecology. Governmental responses to present demands may prove both ecologically tolerable and politically sufficient. But what if policy-makers and citizens increasingly perceive a more severe pollution problem or unacceptable deterioration in environmental quality, despite changes which result from the confrontation of environmental concerns and economic expansion? What if compromises achieved by bargaining or balancing pollution control with unlimited expansion are insufficient? If so, then we can expect a trend into a political economy of ecology, when resources available for pollution control increase and when environmental quality criteria become predominant in policy-making (more important than traditional economic criteria such as growth and employment levels).

Politics in an era of "ecology" takes on a new form. There is a major *redistribution* of power giving environmentalist forces advantages over economic interests.[20] Indeed, the interests of future generations receive consideration through planning, less immediately self-interested policies, and new rates for "discounting" the future. Traditional, territorial political boundaries give way to more regional solutions geared to the "problem shed." International solutions and cooperation may in many areas be the sole acceptable approach, going well beyond the national solutions of previous eras. Political parties, formed around environmental goals, may

19. Dean Schooler, Jr., *Science, Scientists and Public Policy* (New York: The Free Press, 1971), pp. 137–165.

20. Thomas Vitullo-Martin, "Pollution Control Laws: The Politics of Radical Change," in L. Roos, Jr., *The Politics of Ecosuicide* (New York: Holt, Rinehart and Winston, 1971), pp. 346–367. This effort by Vitullo-Martin discusses such a reallocation of national values and resources toward environmental quality.

become significant factors in elections or government policy-making. Policy-making takes on a fully centralized character, since choices are no longer compartmentalized into specific pollutants (media problems such as air and water pollution, substantive problems such as pesticides and toxic materials) or separated decision processes for economic decisions and environmental decisions. Policies would be effectively established by the executive branch or administrative arm of government.

These new conditions which give rise to a politics of ecology stem from a heightened severity of the environmental problem. No longer could anyone contend that the "crisis" was manufacturered and ecologically premature. Not only would the recent concern with renewable biological resources (pollution) continue, but the older preoccupation with a scarcity of non-renewable material resources would rise again to importance. This two-pronged concern has only recently appeared in the area of electrical energy needs and environmental deterioration. Not only are societies concerned about pollution from existing sources of electricity (coal and nuclear power), but some see limits on available fuel resources (uranium, low sulfur coal). These fears have led to research and development efforts in new technologies such as fusion, liquid-metal fast breeder systems, and solar energy.

The severity of the environmental problem thus pushes elites and key national leaders to forge and impose a comprehensive, integrated solution. Decentralized, non-synoptic, fragmented and remedial actions and programs are deemed insufficient. Central decision-makers take full responsibility, and power further accrues to them inasmuch as the issue is defined in "system-wide" terms rather than as several isolated problem areas.[21] Coercion through the regulation of both individuals' conduct and the environment of their conduct becomes essential and acceptable.[22] Citizens' willingness to jeopardize their own personal interests by agreeing to a communal, centrally-determined solution rises. The demand for decisive action over-rides individuals' need to have a voice in their own fate, and an ethic based on co-operation rather than conflict may become operative. Environmental quality takes on a "collective, communal" character and, as a result, environmental policies and programs seem to be a non-disaggregable, shared, communal benefit (cf. defense policies and programs). The game thus is transformed

21. Allen Schick, "Systems Politics and Systems Budgeting," *Public Administration Review*, XXIX:2 (March-April, 1969), 137–151. Schick discusses the advantages of a "systems" approach to policy-making for policy-makers in "mixed mobilizing-rationing" and "central allocation" roles.

22. Lowi, "Decision-Making vs. Policy-Making..," 314–325. Coercion which directly affects an individual's conduct in a specific situation should be distinguished from coercion which operates by changing the general environment which indirectly and more subtlely affects his behavior along with others.

into a *positive non-zero sum situation*, having just been a conflict-based *zero sum situation* that threatened to become a *negative non-zero sum game* as environmental quality deteriorated further. As a result, the objective of "expansion and resource development" or reliance on a solution through the presence of "environmental confrontation" (conflict) are both subjected to serious limitations. Neither "expansion" nor compromised, bargained solutions to conflict have proved sufficient, and this new era has emerged.

The "politics of ecology" involves several specific efforts through governmental policies and activities. These include:

Major Crash Commitment for Revolutionary Technological Breakthroughs. When existing technologies (internal combustion engines, production of electrical power through fossil fuel burning and nuclear plants) prove environmentally troublesome, policy-makers may commit *vast funds* to "crash" programs of research and development toward revolutionary methods of pollution control, propulsion, and energy production. Support for fusion research, solar energy studies, geothermal surveys, and fast breeder systems, though funded at low levels during the 1960s and 1970s in the United States, would be examples of such revolutionary systems. Ecological necessity, rather than cornering the world market and sales for such systems, would be the "mother of invention."[23] Technologies for recycling materials and using waste products also become R and D objectives.[24]

Development of Large Organizations for Environmental Management. Irvin L. White has argued, working from models of the Manhattan Project and U.S. National Aeronautics and Space Administration, that societies might already during the 1970s consider such large-scale, technology-based forms of project organization regarding environmental quality and energy policy-making. These organizations would be organized around a function (environmental management) and would develop an integrated system complete with real time feedback capacities, monitoring systems and adjustment mechanisms.[25] Possibly such organizations might develop around themselves an "environmental complex" rivaling the power of the vaunted "military-

---

23. While there are existing depletion allowances for oil, uranium and many other minerals, one wonders what the depletion allowance for heavy water (deuterium which would fuel fusion reactors) would be or how it would be determined. The fact that it exists in relatively "unlimited" quantities in sea water might complicate the question. But surely questions like this, reminders of a previous era, would persist in an era of ecology, despite the larger commitment to pollution-free energy.

24. Funds would also flow into research on advanced transmission and distribution systems (superconductivity, etc.) as well as mass transit systems. Antitrust laws might be waived for large scale joint research or production ventures.

25. Irvin L. White, "Energy Policy Making: The Limitations of a Conceptual Model," Paper delivered to the Southwestern Social Science Association, Dallas, Texas, March, 1971 (Mimeographed.).

industrial complex" of the 1960s and early 1970s.

Adoption of a Wide Range of New Policy Instruments. The environmental policy-makers' options will expand with the addition of new techniques for managing the environment. These will include recycling requirements, site controls, required product modification, and effluent fees or so-called "pollution taxes." Some new instruments were already proposed in 1970 by the U.S. Office of Science and Technology for the nations' coming energy policy and its electric power system. These were controls on the siting of power plants and transmission lines, elimination of promotional rates, shifts to new fuels or production processes, use of waste heat for space heating and desalting sea water, inclusion of total environmental costs in the price of electricity (air and water pollution abatement, "land" pollution from transmission lines, R & D costs for development of less-polluting means of production). The environmental management package may also involve reductions in population and population density, zoning to achieve better distributions of polluters, outright bans on selected pollutants, and possibly a constitutional right to a clean environment.

Policy-making and environmental administration in the "ecological" era resemble politics in Allen Schick's "cybernetic state."[26] The key techniques would be effluent fee systems and, possibly, a materials use and balance approach.[27] These approaches fit into a cybernated system because they can be altered periodically in response to changes in the environment. Effluent fees or taxes on polluting substances (lead, sulfur oxides) can be varied to achieve desired levels of environmental quality.[28] Since these fees would be determined administratively (and probably not through negotiation with polluters) scientists, economists and planners would have an important role.

26. Energy Policy Staff, Office of Science and Technology, *Electric Power and the Environment.* (Washington: Government Printing Office, August, 1970).

27. Consult Robert M. Solow, "The Economist's Approach to Pollution and Its Control," *Science*, 6 August 1971, pp. 498–503; Allen V. Kneese, "Strategies for Environmental Management", *Public Policy*, XIX:1 (Winter, 1971), 37–52; and E.S. Mills, *User Fees and the Quality of the Environment* (forthcoming); The Second Annual Report of the U.S. Council on Environmental Quality; *Environmental Quality* (Washington: Government Printing Office, 1971) favors increased use of effluent fees.

28. Effluent fees constitute a variation from the earlier standards approach. Standards cannot be manipulated as easily, and though more politically acceptable, are inflexible regarding requirements on an individual polluter. A fee system, coupled with a minimum standards system, would be more "economic" in securing the greatest reduction in effluent for the least expenditure. And it would provide continuing incentive for further reductions. There is a major dissent to this view on the "economic" and "rationality" arguments for administratively-set fees. George Hagevik argues that under conditions of uncertainty, a bargaining process reminiscent of the politics of "environmental confrontation" would be more economic and lead to a more rational solution. Hagevik, "Legislating for Air Quality Management," *Law and Contemporary Problems*, 33:2 (Spring, 1968), 369–398.

Indeed, environmentalists would find themselves attempting to influence the "program" which governs the fee-setting system and controls that system's response to changes in the environment. The debate, as Schick points out, would be similar to deciding who sets the thermostat in a room.[29] Government would write the program, set the norms or minimums, monitor the environment, and collect the fees. The approach might even be extended to the use of all materials and encompass the twin problems of pollution and resource availability. But in either case, policy-making would have moved beyond specific decisions for specific situations and into general rules for whole classes of situations. What was once a debate over specific, substantive regulations would become a debate over procedures and the "program". The politics of ecology would have become cybernated.

(Summary table on following page.)

---

29. Allen Schick, *op. cit.*

Summary table of the three eras of economic development and environment

AREA OF COMPARISON \ ERA	POLITICAL ECONOMY OF EXPANSION AND RESOURCE DEVELOPMENT	POLITICAL ECONOMY OF ENVIRONMENTAL CONFRONTATION	POLITICAL ECONOMY OF ECOLOGY
MAJOR POLITICAL OBJECTIVES	ECONOMIC GROWTH, RESOURCE EXPLOITATION, UNIQUE AREA PRESERVATION	CONTINUED EXPANSION AND ENVIRONMENTAL QUALITY (POLLUTION)	CRISIS, SURVIVAL ENVIRONMENTAL QUALITY (POLLUTION, RESOURCES)
LOCUS OF POLICY-MAKING	LEGISLATIVE COMMITTEES LOW LEVEL ADMINISTRATIVE AGENCIES	ENTIRE LEGISLATURE HIGH LEVEL ADMINISTRATIVE AGENCIES	ADMINISTRATIVE BRANCH
NATURE OF POLICY-MAKING SYSTEM	LOGROLLING, PORK BARREL AD HOC, REMEDIAL SEGMENTED AND SECTORAL DECENTRALIZED LOCAL INITIATIVE AND SOLUTIONS	BARGAINING, COMPROMISE REMEDIAL NATIONAL SOLUTIONS MORE CENTRALIZED GROUP CONFLICT	ADMINISTRATIVE DECISION MAKING, CENTRALIZED COMPREHENSIVE AND INTEGRATED ROUTINIZED, CYBERNATED PROBLEM SHED APPROACH
DOMINANT POLITICAL FORCE	SELF INTERESTED ECONOMIC GROUPS (INDUSTRY, MINING, AGRICULTURE) TRADITIONAL GEOGRAPHICAL AND POLITICAL AREAS	SELF INTERESTED ECONOMIC GROUPS ENVIRONMENTALIST GROUPS	ENVIRONMENTALIST GROUPS (BALANCE SHIFTED BY ELITE INTERVENTION)
TYPE OF GOVERNMENT POLICY	DISTRIBUTIVE SPECIFIC TO SINGLE, ISOLATED AREAS OF ACTIVITY	REGULATIVE SECTORAL BUT MORE UNIFIED	REDISTRIBUTIVE AND REGULATIVE UNIFIED, REGIONAL AND INTERNATIONAL SOLUTION
TYPE OF POLITICAL GAME	POSITIVE NON-ZERO SUM	ZERO SUM	POSITIVE NON-ZERO SUM AS RESULT OF

AMOUNT OF COERCION	MINIMAL	SOME, WORKING THROUGH INDIVIDUAL CONDUCT	MODERATE AMOUNT, WORKING THROUGH INDIVIDUAL AND ENVIRONMENT OF CONDUCT
TECHNIQUES FOR ACHIEVING POLICY OBJECTIVES	SUBSIDIES, TAX CREDITS, DEPRECIATION ALLOWANCES, DEPLETION ALLOWANCES, NUISANCE LAWS	STANDARDS, CRITERIA	MAJOR COMMITMENT OF R&D FUNDS, LARGE SCALE TECHNOLOGICAL PROJECTS, PROJECT FORMS OF ENVIRONMENTAL ADMINISTRATION, EFFLUENT FEES, POLLUTION TAXES WITH MINIMUM STANDARDS
SPECIFIC INTERESTS BENEFITTING FROM GOVERNMENT ACTIVITIES	EXTRACTION AND MANUFACTURING INDUSTRIES, MUNICIPAL SEWAGE SYSTEMS	ENVIRONMENTAL INTERESTS, EXTRACTION AND MANUFACTURING INDUSTRIES, MUNICIPAL SEWAGE SYSTEMS	ENVIRONMENTAL INTERESTS
TYPE OF TECHNOLOGICAL BASE	MINIMAL	POLLUTION CONTROL SYSTEMS	RECYCLING SYSTEMS, REVOLUTIONARY NEW PROPULSION AND ENERGY PRODUCING TECHNOLOGIES
RELATIONSHIP TO ELECTRICAL ENERGY NEEDS AND ENVIRONMENTAL PROBLEMS	UNLIMITED EXPANSION OF CAPACITY; POLICY IN TERMS OF SPECIFIC SECTORS (OIL, COAL ETC.)	MOVEMENT TOWARD AN "ENERGY POLICY" INCREASING GOVERNMENT REGULATION	CONVERSION TO REVOLUTIONARY TECHNOLOGIES, LIMITS ON GROWTH AND SITING; COMBINED ENERGY AND ENVIRONMENT POLICY

Chapter III

ENERGY, ENVIRONMENT,
AND INTERNATIONAL INTEGRATION*

by *Francis A. Beer*

The purpose of this essay is to offer some rather broad speculation concerning the relationship of energy, the environment, and international integration. In so doing it will discuss generally the political and economic relationships between the sectors of energy and environment and between various levels of political systems; the past history of international cooperation in energy and environment; strategies for future international cooperation; and, finally, the likelihood of increasing international integration.

Competitive Sectors

Recent public attention to environmental issues implies that energy and environment are competitive sectors. We are told that we must choose whether we are to have more energy, or even the present amount, or whether we are to preserve what remains of man's natural habitat.

Energy production and environmental preservation seem to exist in a relationship which is almost zero-sum. If a dam is built to generate hydraulic power, it may open new lands for cultivation, but it will inundate others and change the pattern of soil deposit. If coal, gas, or oil are extracted from the earth, the world's reservoir of fossil debris is that much smaller. If facilities are built for the generation of power through processes of nuclear fission or fusion, there are problems of radioactive wastes and dangers of radioactive leaks. The burning of fossil fuels releases circulating atmospheric wastes, while the operation of dams and nuclear plants may change hydraulic levels, currents, and thermal patterns.[1]

* I am indebted to Dean Schooler, Jr. for some of the references in this chapter, as well as for insights which evolved through extended discussion.

1. See, for example, Philip H. Abelson, "Costs versus Benefits of Increased Electric Power," *Science*, No. 3963 (December 11, 1970), p. 1159; Donald F. Anthorp, "Environmental Side Effects of Energy Production," *Bulletin of Atomic Scientists*," Vol. XXVI, No. 8 (October, 1970), 39–41; Paul Jacobs, "Precautions Are Being Taken by Those Who Know: An Inquiry into the Power and Responsibilities of the AEC," *The Atlantic* (1971);

The interests of each sector seem to be supported by competitive political groups with not only different substantive foci, but also different structural characteristics. In general, energy-oriented political actors appear to be closer to the centers of political power than their environmental counterparts. The spirited and long-standing defense of coal, oil, gas, and timber interests in the United States, and the burgeoning atomic power bureaucracy—against the usually Pyrrhic opposition of conservation groups—illustrate the long standing alliance between natural and political forms of power.

This pattern of differential political access is probably related to a number of other apparent differences. Ideologically, the energy groups are on firmer ground within the context of conventional Western democratic theory. By emphasizing the sanctity of property, the benefits of free enterprise, and the limits of state activity, they achieve an advantage over their opponents who call for increased governmental incursion into the private sector.[2]

Organizationally, energy groups can rely upon more stable bases of support. Financial resources are available as a result of past successes; environmental groups are usually strapped for cash. The corporate membership of energy interests is relatively permanent, while environmentalists come and go with the surge of public sentiment. The organizational structure of energy is relatively well defined, while environmentalists converge and dissemble in more amorphous patterns.

Stylistically, energy groups also tend to be more congruent with inherited political patterns. Oriented toward economic and technological progress, with relatively concrete, obvious, tangible, limited goals and long-range programs, they can appeal to a strong tradition of pragmatism. Environmental groups, on the other hand, are less concerned with the benefits of progress than its costs. They may focus on general, wide problems with long-range implications, but call for immediate solution. Rather than emphasizing

"Energy Crisis: Environmental Issue Exacerbates Power Supply Problem," *Science*, (June 26, 1970), p. 1554; Kenneth Mellanby, "Can Britain Afford to Be Clean?" *New Scientist*, Vol. XCIII, No. 668 (September 25, 1969), 648–650; Dean E. Abrahamson, *Environmental Cost of Electric Power* (New York: Scientists' Institute for Public Information, 1970); M. T. Farvard and J. P. Milton, eds., *The Careless Technology: Ecology and International Development* (National History Press, forthcoming); Joint Committee on Atomic Energy, *Environmental Effects of Producing Electric Power, Hearings* (Washington: GPO, 1969–1970).

2. The validity of the ideological defense is not usually hampered by the reality of various types of governmental assistance to such interests. In the case of government operation of certain sectors, such as the Tennessee Valley Authority and various atomic facilities, energy groups call for transfer of governmental resources to private entreprise. See Aaron Wildavsky, *Dixon-Yates: A Study in Power Politics* (New Haven: Yale University Press, 1952); Norman Wengert, "The Ideological Basis of Conservation and Natural Resources Policies and Programs," *Annals of Science* (November, 1962), 65–75.

incremental problem-solving, they may seem to be more concerned with total social overhaul. Very often they are content merely to define the problem, without necessarily being able to indicate the immediate answer. Their vision, in short, may be less pragmatic than chiliastic or apocalyptic.

A number of commentators argue that the competition between energy and environment is more apparent than real, that relations between energy and environment are much more complicated than can adequately be described through a simple competitive schema, that supposed conflict of interest can be overcome through the development of potential cooperative interest areas, and that stronger political institutions are required to do the job.[3] Thus Daniel Berg concurs with John N. Nassikas, Chairman of the Federal Power Commission, concerning the need for "a comprehensive energy policy to effect balanced objectives of efficient utilization of our energy resources in harmony with the environment." Obviously coordinative political institutions can help to achieve these objectives by retarding the waste of natural resources and providing leadership for their protection.[4]

Opposed Regions

To say that political structures may help resolve apparent conflict between energy and environment, and help to realize potentially cooperative solutions, leads immediately to the problem of the proper scope of such institutional coordination, which can take place at local, national, or international levels.

In general, broader structures are presumeably able to coordinate narrower ones. Thus the problems arising out of various local and regional differences are often supposed to have national solutions. Theoretically, national structures should increase the over-all rationality of policy, the balance of energy and environmental concerns, within national boundaries and overcome the dysfunctional biases of interest groups with local roots, perspectives, and organizational foundations.

At the present time, however, national coordination of energy and environment in many states seems more likely to be myth than reality. The unhappy history of federal regulatory agencies in the United States has

3. See Mancur Olson, Jr., *The Logic of Collective Action: Public Goods and the Theory of Groups* (New York: Shocken, 1965), on the uses of political organizations and institutions in converting sub-optimal to optimal solutions. The role of such institutions is analogous to that of communication in resolving the paradox of the game of prisoner's dilemma.

4. Daniel Berg, "Energy Without Pollution," *Science*, Vol. 170 (October 2, 1970), 17. Energy Policy Staff, Office of Science and Technology, *Electric Power and the Environment* (Washington: GPO, August, 1970) argues in the same vein. Interestingly, Nassikas, with a small group of other "key executives," is formally associated with the report, p. iv.

shown that such structures tend to become lobbyists for producers versus consumers, and the role of mediator is betrayed for that of clandestine advocate.

The promise of international institutions, because of their wider breadth, is even greater than that of national structures; but the likelihood of achieving it is even less. Coordination of diverse national programs and perspectives could multiply the benefits of scale, but is largely prevented by the very problems which are supposed to be solved. Powerful national interests and mythology, together with differential distribution of resources, stand in the way of easy progress and the cheap attainment of optimal solutions.

Between nations, disproportional costs and benefits create barriers to cooperation. Thus industrial wastes in Japan poison swordfish which are caught and eaten by Americans. Conversely, American atmospheric testing of high energy nuclear devices may produce fall-out harmful to Japanese.[5] Moreover, there are probably repetitive differences between more and less economically developed regions. It seems that a high level and rate of economic development in today's world is directly related to both high levels of energy consumption and environmental pollution. In this situation the more developed countries may import energy benefits and export part of the ecological costs.

Asymmetrical International Cooperation

International cooperation has occurred, but it has been asymmetrical in the sense that energy seems to have been more relevant to international integration than environment. This is understandable if we take a standard list of explanatory indicators of integration. It is obvious that energy is required for all forms of international transaction—transportation for products and people, electricity for communication, power for cooperative production projects, etc.[6]

In the more economically developed world, the sector of energy gave primary impetus to the formation of the European Six, in the form of the European Coal and Steel and Community, and provided a second stage boost in the guise of EURATOM. Power from coal and atomic sources

5. See Study of Critical Environmental Problems, *Man's Impact on the Global Environment: Assessment and Recommendations for Action* (Cambridge, Mass.: MIT Press, 1970); Fred Wheeler, "The Global Village Pump," *New Scientist*, Vol. 48, No. 721 (October 1, 1970), 10–13; Reginald E. Newell," The Global Circulation of Atmospheric Pollutants," *Scientific American*, Vol. 224, No. i (January, 1971), 32–42.
6. A good summary of the current state of international transaction analysis is presented by Donald Puchala, "International Transactions and Regional Integration," *International Organization*, Vol. 24, No. 4 (Autumn, 1970), 732–763.

seemed attractive as a nucleus for international cooperation. At the universal level, the International Atomic Energy Agency represented a similar structure. Many smaller international organizations, both public and private, have also been concerned with energy. Particularly in the field of petroleum, a large and complex network of international corporations and the Organization of Petroleum Exporting Countries have been major actors.

In the developing areas, the supply of energy is one of the few really crucial bargaining levers against the more developed world. Those states that have oil resources are beginning to discover the virtues of cooperative international action against the consumer nations. The result of this should be not only to increase their financial resources, as in the oil crisis of spring 1971, but quite possibly to give impetus to efforts to form limited regional communities. Integrative learning is not always transferred between sectors, but the autonomy of sectors is not absolute. If nations can achieve income creation through joint efforts, it seems reasonable to believe that a small part of the new wealth may be used to shore up common institutions—for example, the League of Arab States, the Regional Cooperation for Development (Pakistan, Iran, Turkey), the smaller Machrek (United Arab Republic, Sudan, Libya), and the Maghreb (Tunisia, Algeria, Morocco).

On a larger scale, the supply of energy should prove of some bargaining efficacy within the framework of universal international organizations. Thus, although the developing nations have had little influence, they have managed to form an articulate and active coalition in UNCTAD for the international redistribution of wealth. The possession of scarce resources by a significant number of UNCTAD's less developed members might well make their arguments more persuasive as time goes on, if only for particular cases.

At the same time, disintegrative forces are at work. A decline in the relative importance of coal as an energy source implies the diminished importance of this part of ECSC activity; and the high political stakes of atomic energy production have stood in the way of substantial contributions to European community building so far.[7] In the developing areas, the leverage of oil may represent only a temporary phenomenon. The prospect of cheap and easily available solar energy, which might particularly benefit the developing nations, does not seem to be for the immediate future; but energy from nuclear fusion may be almost within reach. If it does become available, the

7. J. Guéron describes the accomplishments and problems of EURATOM, CERN, ENEA, and other cooperative international nuclear energy projects, with particular emphasis on national preoccupation with "fair return" and "slow return", in "The Lack of Scientific Planning in Europe," *Bulletin of the Atomic Scientists* (October, 1969), 10–14, ff; "The Lessons to be Learned from Euratom," *Bulletin of the Atomic Scientists* (March, 1967), 38–41; "Atomic Energy in Continental Western Europe," *Bulletin of the Atomic Scientists* (June, 1970), 61–68, ff.

more developed nations will have an alternative source of power.

In environmental matters, the international record is weaker. One looks in vain for important international organizations with environment as the core activity. Perhaps the most substantial international environmental accomplishment is the partial nuclear test ban. Organizationally most impressive are the World Meteorological Organization, with the World Weather Watch, and the activities of the International Council of Scientific Unions, leading to the International Geophysical Year, the signature of the Antarctic Treaty, the International Indian Ocean Expedition, the International Year of the Quiet Sun, and the Global Atmospheric Research Program.[8]

Generally the pattern has been to adopt environment as a kind of step-task, in addition to more natural and central progeny. Thus the Organization for European Economic Cooperation, during the years of post-war recovery, studied atmospheric sulphur from the perspective of retrieving a scarce raw material.

In recent years there has been more concentration on environmental problems as worthy of study in their own right, and an accelerated rhythm of international action has appeared. The OECD has established general programs for the study of water, air, urban environment, and toxic chemicals. A number of ad hoc investigations have been started including "air pollution from combustion of fossil fuels, pollution resulting from the generation of electric power, and nuisance aspects of the automobile." The United Nations and its various segments, the Council for Mutual Economic Assistance, the Organization of American States, the Council of Europe, the European Communities, and even the North Atlantic Treaty Organization have all devoted resources.[9]

Integrative Strategies

In order to relieve sectoral conflicts, resolve regional oppositions, and balance international asymmetries, an integrative strategy, emphasizing international concern for environmental effects of energy production, would seem to be desirable.

8. Ernst B. Haas, "Science, Technology, and Planning," (Berkeley: mimeo), p. 9/9.
9. John A. Haines, "International Cooperation" (Paris: OECD mimeo, June, 1971), p. 3. See also M. Nicholson, *The Environmental Revolution: A Guide for the New Masters of the World* (London: Hodder and Stoughton, 1967), Ch. 9: United Nations Economic and Social Council, 47th Session, "Problems of the Human Environment," Report of the Secretary General, Doc. E/4667, May 26, 1969; "The United Nations and the Environment," *American Journal of International Law*, Vol. 64, No. 4 September, 1970), 211–238; Leon Gordenker, ed., *The United Nations: Its Future, the World Environment, and U.S. Policy* (forthcoming); *Utilisation et Conservation de la Biosphère* (Paris: UNESCO, 1970); A. Szalai, "Behind the Alarm," *FAO Review* (May-June, 1970); European Information

Existing theory and research on integrative processes indicates that two crucial levers for producing integration may be leadership and crisis.[10] Leadership may come in either national or international contexts. Nevertheless, it would probably be beneficial for environment to have its own international spokesmen.

In particular, it should be possible to establish an International Environmental Agency, within the United Nations family, to act as "an organizational personality—part conscience, part voice, which has at heart the interests of no nation, no group of nations, no armed force, no political movement, and no commercial concern, but those of mankind generally, together with man's animal and vegetable companions who have no other advocate."[11]

Minimal tasks for the IEA might include the collection, storage and retrieval of information; coordination of national research and operational activities; promulgation of international standards and advice; and establishment of rules for human activities on the high seas and in the stratosphere, outer space, the Arctic and Antarctic.[12]

A more ambitious conception would involve independent planning, funding and operations; monitoring and enforcement of standards and rules; mediation, conciliation, arbitration, and adjudication of disputes.[13]

The leadership's ideology should aim at either co-opting or neutralizing part of the energy opposition. To some extent this should be possible by spreading resources such as governmental contracts and grants. Cost-benefit system studies can be undertaken, focusing on maximal environmental purification with minimum energy dislocation. Much work is already under way in this area; but there exists the danger that the environmentalists will end up

---

Centre for Nature Conservation, *Council of Europe Documents about the Environment* (Strasboug: Council of Europe, May, 1971); J. Tinker, "ECY: Steering the Bandwagon," *New Scientist* (February 5, 1970); Commission of the European Communities, Directorate General for Industrial, Technological and Scientific Affairs, "Electrical Energy Needs and Environmental Problems in the Community" (Brussels: mimeo, June, 1971); "Rapport sur la Lutte contre la Pollution des Eaux Fluviales et Notamment des Eaux du Rhin," Document 161, Parlement Européen, Commission des Affaires Sociales et de la Santé Publique (November 11, 1971); "Action: Key Word of the CCMS," *NATO Letter*, Vol. 18, No. 2 (December, 1970), 9–13.

10. Francis A. Beer, *Integration and Disintegration in NATO* (Columbus: Ohio State University Press, 1969), Ch. 7.

11. George F. Kennan, "To Prevent a World Wasteland," *Foreign Affairs*, Vol. 48, No. 3 (April, 1970), 401–413. See also Abel Wolman "Pollution as an International Issue," *Foreign Affairs*, Vol. 47, No. 1 (October, 1968), 164–175.

12. *Ibid.*

13. Eugene B. Skolnikoff, "The International Political Imperatives of Science and Technology," Paper prepared for presentation at the Seventh Edinburgh Seminar in the Social Sciences, May 25, 1970, pp. 62–65.

being the co-opted rather than the co-opters. Energy production will continue, but environmental pollution may be relatively unchecked.[14]

To prevent environmental programs from turning into environmental rackets, it would be well to create strong agencies at both national and international levels, with substantial budgets and personnel complements, with responsibility for environmental matters. Large scale training and degree programs in the environmental sciences might be established to create a new corps of technical environmental experts and administrators.

One important function of such agencies would be to promote publicity for environmental crises. A psychological phenomenon beginning to receive significant attention is what may be called "the obscurity of the obvious." In order to make the obvious obvious, and in order to give the environment relevance for political action, some element of crisis is probably required. The yellow-orange smog of Los Angeles, the floating bodies of tons of dead fish on European rivers, contamination of whole species of ocean fish, are beginnings in this direction. Such crises are not to be desired. If one would follow the logic of catastrophic optimism all the way to its logical conclusion, widescale irreversible disaster might be the result. Nevertheless, they have symbolic uses. They serve to dramatize, to provide a concrete form of communication, a common body of primary experience, which may help to change the general perceptions of large bodies of individuals.[15]

From what we have seen in the last few years, we may not lack such crises in the decade to come. One would almost be tempted to predict that the 1970's would be the decade of natural disaster, were it not for the fact that, sooner or later, the educational effects of such happenings may be expected to take, and wide-spread political movements may form the basis for palliatory action. Already in the field of automobile pollution, there are signs that technology may cure some of the ills which it has created. New and more effective filtering devices for hydrocarbon emissions, together with alternative steam and electric power plants, may largely solve this problem. Although the advent of fusion sources for electric power may well replace one sort of environmental pollution with others, it promises to solve at least some of the existing maladies.

It is also possible that new international cooperation will come with new

14. For a classic study of reverse co-optation see Philip-Selznick, *TVA and the Grass Roots* (Berkeley: University of California Press, 1953). James Ridgeway in *The Politics of Ecology* (New York: Dutton, 1971) describes the extent to which this process is already well advanced in the contemporary ecology movement. *Energy, Economics, and the Environment* (Chicago: Committee on Environment, Edison Electric Institute, 1968) is a good example of how it is done.

15. See R. D. Laing, "The Obvious," in David Cooper, ed., *The Dialectics of Liberation* (Middlesex: Penguin, 1968), 13–33 and Murray Edelman, *The Symbolic Uses of Politics* (Urbana: University of Illinois Press, 1967).

measures of environmental control. Some of the cooperation will be fortuitous. Thus, the adoption of emission control standards for automobiles in the United States may force manufacturers exporting there to tool up for the American market, and the results may then spill back into the domestic economies. National governments may move to make smog control devices mandatory in response to pressures both from environmental and industrial groups. Other forms of cooperation may involve more formal kinds of institutional coordination, for example international treaties with regard to river pollution in Europe. Nevertheless, national environmental control standards and devices may also serve as the leading edge of a new and more sophisticated generation of non-tariff barriers, and thus help to undermine efforts at international economic cooperation.

Much environmental destruction, however, not yet at the crisis stage and without immediately observable concrete effects, may go largely unchecked by either national authority or international agreement. Green belt forests in Africa and Latin America, supplying a large percentage of the world's oxygen, may be reduced to meet locally perceived needs for energy and development. Waste disposal in ocean spaces is likely to continue for some time. New fishing processes may help to de-populate the seas with a rapidity reminiscent of the disappearance of the buffalo from the American plains. High protein fish flour offers the possibility of nourishing large numbers of people, but with possibly irreversible effects on the source of supply.

The Limits of Integration

It is difficult to assign a high probability to energy or environmental crises which would quickly produce international integration of the two sectors. Some analysts believe that science and technology in general present "political imperatives" for integration.[16]

In this connection it would seem well to remember that the new technology can work not only to create greater international cooperation, but also international conflict.[17] Scientists can work and have worked within national boundaries in deepest secrecy for strictly national ends. Moreover, different scientific groups may have different political and economic interests and

16. Skolnikoff, "The International Political Imperatives of Science and Technology."

17. Suggestive analyses appear in *The Potential Impact of Science and Technology on Future U.S. Foreign Policy*, Papers presented at a joint meeting of the Policy Planning Council, Department of State and a Special Panel of the Committee on Science and Public Policy, National Academy of Science, June 16–17, 1968, Washington, D.C.; N. Calder, *Technopolis: Social Control of the Uses of Science* (London: Macgibbon and Kee, 1969), Ch.9; Rita F. Taubenfeld and Howard J. Taubenfeld, "Some International Implications of Weather Modification Activities," *International Organization*, Vol. 23, No. 4 (1969).

access. In connection with the present analysis, it has seemed that natural and earth scientists have been more interested in energy, while life scientists have more often concentrated on the environment.

It is not at all impossible that confrontations between national energy suppliers and consumers will lead to energy wars. The semi-permanent conflict in the Middle East contains elements of one such scenario. The external costs of various forms of atmospheric or ocean pollution, manipulation, or destruction may, in similar fashion, lead to wars over the environment.

In order for the "political imperatives" of energy and the environment to be realized, they must first be recognized by the relevant political actors. The lessons of international functionalism over the last generation would indicate that a certain amount of caution may be in order. Paradoxically, international organizations share many of the problems of the environments which they might protect. Rather than being effective actors on the stage of international politics, most often they represent the stage and provide the scenery for the drama which is to be played. In the same ways that processes of national and sub-national governmental regulation may contribute to the despoliation of man's natural resources, the self-preoccupation of national political elites contributes to the continuing subordinate role of international organizations. In the same way that actions upon the environment are disconnected and relatively uncoordinated, actions in the international arena also occur as part of a multidimensional pattern.

If there is progress in international cooperation, it is likely to be piecemeal and slow, incremental rather than radical. As has been the case in the past, existing institutions may take an interest in the relationship between energy and the environment. Some progress may be made toward authoritative decision and regulation, as in existing European agreements on ionization levels and in the action program being prepared by the European Communities. This program involves "the common use of criteria, indices, and standards, the establishment at the Community level of bodies for environmental inspection and management, the implementation of research programs and various specific schemes," either on geographical or industrial bases. New institutions and leadership positions may be established to give increased voice to environmental concerns. But in most cases such institutions are not likely to move rapidly toward effective actions which may substantially redistribute costs and benefits, within or between nations. Stronger states will probably defend their particular interests as vigorously as before against the claims of the weak; energy advocates will seek, successfully in many cases, to subvert environmental control activities. At the risk of being pessimistic, it is hard to imagine anything much less than imminent world-wide ecological cataclysm which might substantially accelerate this glacial pace of change.

Part II

TECHNICAL PROBLEMS
AND NATIONAL ENERGY POLICIES

Chapter IV

ELECTRICAL ENERGY NEEDS AND ENVIRONMENTAL
PROBLEMS, NOW AND IN THE FUTURE*

Introduction

Electricity consumption in the industrialized nations has been doubling every ten years for several decades. This rate of growth has been caused by an increasing need for electricity in domestic, public and industrial spheres, even apart from the influence of the population increase. There are, as yet, no indications of a slackening of growth in the demand for this convenient and—at the point of consumption—clean form of energy.

Extensions of energy supply systems have to meet economic criteria. They must be reliable and safe, and may not affect unduly the environment. In view of the requirements imposed this publication examines today's technological and economical solutions and discusses the feasibility of propositions for future developments.

The long lead times that go with research, development and manufacture, together with an acceptable technical or economic lifetime of the installations make it appropriate to cover the next three to four decades in this survey.

Many examples used in this study are based on the situation in the Netherlands, the country with the highest power density (MW per km²) in the world. The analysis has, however, been made in such a way as to make it of more than just national importance.

* Condensed version of publication number 7 under the same title of the FUTURE SHAPE OF TECHNOLOGY FOUNDATION, THE HAGUE, THE NETHERLANDS, APRIL 1971. About this Foundation, see Foreword.

ELECTRICAL ENERGY, DEMAND AND SUPPLY

by *J. H. Bakker* and *J. J. Went*

Summary

The demand for electrical energy is increasing at an even faster rate than the growth of energy consumption in general. To satisfy the demand, electric power is produced mainly in thermal power stations, using fossil or nuclear fuel.

This kind of production is attended by the heating of cooling water and —for fossil fuel plants—by air pollution. Production units placed outside heavily populated areas allow better control of harmful environmental effects than plants using direct combustion of fossil fuel within cities.

Electricity enables clean use of energy where it is needed. These facts may lead to a still larger growth rate of electricity consumption.

For future electric power production breeder and fusion reactors are evaluated and various direct conversion methods are discussed.

Present-day and future transmission systems are analyzed.

General Remarks on Energy

The sun is the energy source that was, is and will be sufficient to make plant and animal life possible on earth for a practically infinite future. However, our civilization requires additional concentrated energy for many purposes. This amount of energy is extremely small compared with the energy we are receiving from the sun. Thermal solar energy intercepted by the earth is about 100.000 times greater than present installed electrical capacity. Energy consumption is increasing so fast, however, that there is a real problem with respect to the availability of this additional energy source and to the environmental aspects.

Electrical Energy

It is possible for electricity to meet man's requirements for power and heat as well as for light and telecommunication. For telecommunication electricity is the only possible energy source, and for lighting it is by far the most efficient and simple. For mechanical power in general, and for all mass transport (except shipping), it is equally important. And yet, 75% of all energy needs are still satisfied without the intermediate link of electricity. As the growth rate of electrical energy consumption (7–12% per year) has been twice the growth rate of energy consumption in general (3–5% per year) for a very long time, it is to be expected that the gradual increase in electricity's share of the market will eventually reach at least 50%.

There is yet another very important reason why electrical energy production should increase, particularly in highly developed parts of the world. When *applying* electrical energy, it is clean, not disturbing or poisoning the immediate surroundings of the place of application. When *producing* electrical energy, this is not always the case.

In general, electric power is generated in large units in central power stations. Apart from hydroelectric power production, which is of some importance in the present world, most electric power is produced in thermal power stations.

In these thermal power stations either fossil or nuclear fuel is used. With fossil fuel environmental effects from combustion products may occur in the atmosphere in addition to the heating of cooling water. With nuclear fuel in nuclear fission reactors, radioactive fission products are formed. These fission products, although contained, can be a potential hazard to the surroundings. However, their quantity is so small compared to the amount of combustion products (a factor of one over 20 million) that their concentration and storage is technically possible. Therefore atmospheric pollution can be avoided with nuclear reactors.

The heating of cooling water, however, is of the same order of magnitude for either fissile or fossil fuels. In modern light water nuclear steam plants the amount of cooling water required is about 1.7 times greater than in modern fossil fuel steam plants.

The Current Demand for Electrical Energy

The demand for electrical energy is a function of the growth of the gross national product. In highly developed countries such as the United States, Canada and Sweden the demand is very high (about 8000 kWh/year/inhabitant). In the Western European countries the demand lags behind by about 10 to 15 years, consumption being about 3000 kWh/year/inhabitant. In the underdeveloped countries the electrical energy demand is still very low.

In assessing the environmental effects of electrical energy production, it is more important to know the electricity production per km^2 than per inhabitant.

Additional important factors are the presence of a sufficient percentage of water surface per km^2 for cooling purposes, and the strength and directions of prevailing winds for evaluating the air pollution. In Table 1 data are given for 1968 for different European countries, for the EEC as a whole, and for the United States. The countries are listed according to their thermal electricity production per km². It is immediately clear from this table that the population density is the most important factor determining the production figure per km². For all countries two figures are given: the thermal electricity production in GWh/km²; and the total electrical energy consumption in GWh/km².

The latter figures are higher in the countries with hydroelectric energy production and are also slightly different from the first figures due to some import or export of electricity.

The future demand for Electrical Energy

Energy demand in the highly developed countries is and has been growing for many decades at a rate of 3–5% per year. Electrical energy consumption, however, is growing much faster: 7–12% per year, resulting in doubling times of roughly 10 years. These doubling times have been constant for about half a century, and apply to a country such as the Netherlands with an intermediate high consumption level (about 2500 kWh/capita) as well as to the United States or Canada with about 7500 kWh/capita. The important question to be answered is: "Just how probable is it that this trend will continue in the years to come?"
There are good reasons to believe that it will.
   One reason for this trend might be the fact that the use of electrical energy is a completely clean application of energy at the place where it is used. Nuclear energy production instead of fossil energy production would also reduce air pollution considerably. The availability of fuel reserves, is unlikely to play an important role in the next few decades but will undoubtedly be very important in the distant future. The fact is that the available fossil fuel reserves are diminishing so fast that a future without fossil fuel must be considered. The application of nuclear energy with the transformation of nuclear energy into electrical energy seems to be the best way of supplying energy where it is needed. As a general conclusion it seems reasonable to assume an average doubling time of 10 years in the Netherlands for the next three decades.

Transmission Systems

In order to satisfy demand, electricity has to be transported to the centres of consumption. In addition a system is required for the exchange of energy between different production systems for reasons of safety and economy. However, electricity transport is very expensive compared with transportion

Table 1. Electric power density in Western European countries and in the United States (1968)

country	population	area	population density	total electricity consumption			thermal electricity production	
	× 1000	km²	inhabitants/km²	GWh	kWh/inhab.	GWh/km²	GWh	GWh/km²
Netherlands	12.750	33.500	380	31.500	2470	0,94	31.800	0,95
Belgium	9.619	30.500	320	23.000	2400	0,75	24.700	0,81
United Kingdom	55.391	244.000	227	215.000	3880	0,88	181.000	0,74
Germany	60.205	248.500	242	197.000	3270	0,79	171.500	0,69
Italy	53.211	303.800	175	102.000	1920	0,34	53.400	0,18
France	50.082	551.200	91	120.000	2400	0,22	64.400	0,12
E.E.C.	185.867	1.167.500	159	473.500	2550	0,41	345.800	0,30
U.S.A.	201.166	9.363.400	21	1.435.000	7140	0,15	1.207.000	0,13

of gas or oil. Therefore new developments are required, to find better solutions. Whereas overhead transmission lines meet more and more opposition and difficulties in populated areas and a switchover to present-day cables would greatly increase the price of electricity transport, it would be desirable to evaluate new technological possibilities such as the development of cryogenic cables, perhaps in a superconducting state. It seems unlikely, however, that such a development will be completed within the next decade. Another solution to reduce the required transmission capacity might be the building of peaktime units (e.g. gas turbines) in the consumption centres.

Production Methods

The potential amount of water power available in the world, as yet only partially used, is about equal to all the electrical capacity installed at present. Due to the fact that only Africa, South America and South East Asia still have large water power reserves available, these reserves cannot meet future electricity demands in the rest of the world.

The remaining modern production methods are all of a thermal nature. Expansion with thermal power stations, using either fossil or fissile fuels will therefore be dominant in the years to come.

In the thermal power stations either a steam cycle (Rankine cycle) with a steam turbine, or a hot gas cycle (Brayton cycle) with a gas turbine is used to drive the electrical generator. The theoretically possible thermodynamic efficiency of these cycles is determined by the relation $\dfrac{T_1 - T_2}{T_1}$, deduced from the second law of thermodynamics.

T_1 and T_2 are the highest and lowest temperatures of the steam or gas cycle.

The actual efficiency is in fact, owing to various losses, smaller than $\dfrac{T_1 - T_2}{T_1}$.

The highest practical temperature for the steam cycle is at the moment about $540°$ C, and for the open gas cycle a few hundred degrees higher. These temperature limits are connected with the lack of good high temperature materials. The lowest temperatures for the steam cycle are obtained in the condenser of the steam turbine and, for the gas cycle, in the outlet of the gas turbine. For the steam turbines condenser temperatures should not be higher than 30–$35°$ C, as an increase of $1°$ C in the condenser means a decrease in efficiency of about $\frac{1}{4}\%$. For the once through gas turbines higher inlet temperatures are possible, and therefore higher outlet temperatures of a few $100°$ C are still acceptable. Besides, due to the fact that gas turbines are mainly used as peaking units, the thermal efficiency is not the most important cost figure in their operation. As these power stations are so well known, it seems unnecessary to describe them in any more detail.

In present-day light water nuclear reactors the steam temperature must not exceed 300–320° C so as to not approach the critical temperature of water (374° C). This results in lower thermal efficiencies (and, in turn, larger cooling water requirements) for nuclear than for conventional fossil power stations with steam turbines. For advanced reactors with high temperature gas cooling, better efficiencies can be obtained. Furthermore, with helium cooling in sight, a closed turbine loop will be possible.

In Table 2. data concerning these efficiencies and losses are given.

Table 2. Thermal efficiencies and heat losses in different power plants.

type of plant	thermal efficiency	heat outlet	
		to stack	to cooling water
fossil fuel steam turbine plant	38–40%	10%	50%
fossil fuel gas turbine plant	22–25%	75%	no
light water reactor plant	30–32%	no	70%
advanced high temperature reactor plant	38–42%	no	60%

Future Electric Power Production

Consideration should also be given to other electric power producing methods which are under examination at the moment but not yet available for practical application. On the one hand there is fissile fuel, thorium or uranium, in breeder reactors; on the other hand fusion fuel—fusion of hydrogen—in thermonuclear reactors. Both of these, together with direct conversion methods, are evaluated below.

The nuclear reactors discussed before are all convertor reactors, which are consuming the scanty, cheap reserves of uranium 235 at an alarming rate. By the end of the century scarcity of cheap natural uranium will increase nuclear fuel costs considerably. Therefore breeder reactors—whose main requirements are uranium 238 and thorium, both available in abundance— are essential for the large-scale longterm production of nuclear energy. Without going into detail, two groups of breeder reactors should be considered.

Fast breeders with liquid sodium cooling are being developed in many different countries (USA, UK, France, Germany, Belgium, Netherlands, USSR). It is not yet certain how much development time will be needed, nor is it known how economical this reactor will be with regard to the effect of large fast neutron fluxes on the construction materials inside the reactor. Its thermal efficiency could be in the region of 40%. Other fast breeders, with gas cooling, are also being discussed.

Thermal breeders (USA—molten salt reactor, Netherlands— liquid suspension reactor) are another type of production capacity being considered by some scientists. The fundamental material difficulties mentioned for the fast breeders are not so serious for thermal breeders, and, in addition the fissile inventories are much smaller. However, far less development effort is being applied to thermal breeders, so that the development time required will be at least as long as for fast breeders. The thermal efficiency of the suspension reactor will be about equal to the thermal efficiency of light water reactors (30%).

From a physical point of view both groups of reactors are feasible. Just how economically acceptable they will be remains to be seen in the coming decades.

Without a few remarks about fusion reactors this survey would be incomplete. The two fusion reactions that can be discussed are the deuterium-deuterium (D-D) and the deuterium-tritium (D-T) reactions. The first would be the preferred reaction due to the fact that for every 5000 hydrogen atoms, one heavy hydrogen (Deuterium) atom is available in the world, which means an inexhaustible energy source comparable with thorium or uranium 238 in breeder reactors. However, it is possible that only the second reaction, between D and the very heavy hydrogen atom (T), which does not exist in nature, will be attainable due to the required plasma densities and temperatures. As mentioned the amount of deuterium in the world will be quite sufficient even beyond the foreseeable future; the problematical factor is the tritium, which does not exist in nature and therefore must be produced by a nuclear reaction with the element lithium. Since the amount of lithium for fusion energy production is only comparable with the amount of uranium or thorium for fission energy production, it is in fact a technical and economical question which is preferable. In any case fusion cannot be expected to become an energy source in this century.

Direct Conversion Methods for Electricity Production

Apart from the Rankine and Brayton cycles a few other more direct conversion methods for electricity production can be mentioned.

In modern thermal power stations, either conventional or nuclear, the electrical generator is a machine in which a solid conductor is moving in a magnetic field. However, there are other conductor materials, liquid or gaseous, which can also be used. As a liquid conductor, a liquid metal such as molten sodium is a possibility. As a gaseous conductor a very hot ($>$ 2000° C) ionized gas can be used.

Development work is being carried out especially on the gaseous conductors. Magnetohydrodynamic generators using a hot ionized gas are thus being developed. However, such high temperatures are required that satisfactory structural materials both for the gas duct and for the electrodes are not yet available. But, if the material problems were solved, an efficiency up to 70% would be possible. M.H.D. generators using an extremely hot, fully ionized gas, such as is present in a fusion reactor, seem to be science fiction.

The energy of the sun intercepted by the earth is about 100.000 times greater than the electrical capacity installed in the world at the moment. However, transforming this energy by means of photoelectric cells seems to be completely impracticable. With a conversion efficiency of 10% it would be necessary to cover a surface of 40 km^2 for a power station of 1000 MW. And even if this were acceptable, the storage of this energy would be impossible with present methods. The same objection applies to thermoelectric cells.

Thermionic convertors are devices with a very hot electron-emitting cathode and a low temperature anode. Although the thermal efficiency can be reasonably high and the device has the advantage of having no moving parts, the very strict dimensional requirements of the gap between the diodes make a large scale application rather impracticable.

For very small power sources, such as in space vehicles, the last two methods of energy production could be of value. They are now widely used for measuring light intensities and temperature differences. For normal electric power production, however, none of these methods can be considered realistic.

In electrochemical fuel cells chemical energy can be converted into electrical energy on catalytic surfaces. As the surface properties determine the efficient operation of these fuel cells, very clean chemicals such as hydrogen or very pure hydrocarbons are required so as not to poison the catalysts. Although small scale applications may be possible, larger scale use is not foreseen.

Additional Remarks

So far only the production of electricity has been discussed—with a brief look at its transmission. However, there are two cases where heat and electricity may be more advantageously produced simultaneously than separately.

In industries with large low temperature steam requirements it is profitable to produce high temperature steam, using it for the production of electricity by a counter pressure turbine and the remaining low pressure steam for other industrial purposes, such as in paper factories. In these cases the primary need is heat, but relatively cheap electricity can be obtained at the same time. Any further electrical energy required can be supplied from outside. The electricity production per unit of heat input is of course much lower than in normal electric power stations (about 12%). In the Netherlands about 10% of all electricity is produced in this way.

The second case is a "total energy system" where a balance is foreseen between electricity production on the one hand, and central heating or air conditioning on the other without an external supply of one of these energy forms. It seems to be difficult to find a sufficiently constant balance during the year. Even in the United States where the total energy system is strongly advocated, and where air conditioning during summertime is common practice, this type of power production remains a negligible percentage of the total electrical energy production of the country. Furthermore it should be realized that, by using a total energy system in a city, local pollution arising from local energy production from fossil fuels (nuclear reactors are too large in this context) is unavoidable.

Conclusions

Summarizing this short chapter the following conclusions can be drawn:
1. The application of electrical energy is the application of completely clean energy.
2. The production of electrical energy is connected with the following environmental effects:
  - the heating up of cooling water, or the evaporation of water in cooling towers attached to fossil or fissile fuel plants.
  - the pollution of the air by combustion products from fossil fuel plants.
However, by placing electric power production units outside heavily populated areas, environmental effects can be better controlled than in the case of direct combustion of fossil fuel inside populated areas.

DISCHARGE OF WASTE HEAT

by *K. J. Keller*

Summary

In the coming decades electric power will still be produced using steam powered turbines, implicating a need for cooling water.
Demands for cooling water can be met by once through fresh or salt water cooling, by surface cooling or by cooling towers.
The cooling water situation in Western Europe is briefly reviewed. The above-mentioned cooling methods and attendant problems are thoroughly discussed for the Netherlands. Regarding future cooling water needs the available reserves in the Netherlands are estimated.
Some possible beneficial effects of heated cooling water discharge are mentioned.

Introduction

Nowadays electricity is mainly produced in steam cycles[1] and generators are driven by steam powered turbines. For several reasons these steam turbines need condensers with some kind of cooling, e.g.:
. Environmental. Large turbines need hundreds of tons of steam per hour. Discharging such large quantities of steam directly into the atmosphere would cause unacceptable environmental problems.
. Economical. Modern turbines need boiler feed-water of an extremely high purity. It would be very uneconomical to let this highly purified water escape into the atmosphere.
. Thermodynamical. The efficiency of the steam cycle is highly increased by application of a condenser. Steam freely discharged into the atmosphere has to do work against atmospheric pressure, which results in losses. The lower the temperature, the lower the pressure in the (vacuum) condenser and the higher the efficiency.
    The following are alternative methods for electricity production requiring less or no cooling water.
1. Hydroelectric power generation is very attractive. Because neither fuel nor cooling water is needed, there is neither air pollution nor thermal pol-

---

1. J. H. Bakker and J. J. Went, "Electrical Energy, Demand and Supply," *Electrical Energy Needs and Environmental Problems, Now and in the Future*, Future Shape of Technology Publication, No. 7 (The Hague: Future Shape of Technology Foundation, April, 1971), pp. 9–15; pp. 36–44 of this book.

lution. Unfortunately hydroelectric power requires high investment, large height differences and has only a very limited availability.

2. Aeolic power generation has never played a role of any importance. Though recently enormous windmills were planned in Siberia[2] where the "Bora" has a high and constant velocity, this way of power generation will be of even less importance than the previous method.

3. Electric power generation by gas turbines has a very low efficiency and is therefore restricted to special applications such as the generation of peak-power or in combination with other processes where heat is needed. Although gas turbines might play some part in power generation in the future, the cooling water problem will not be much easier.

4. A number of non-conventional ways of power generation such as several methods of direct conversion (thermoelectric, thermionic, magnetohydro-dynamic, fuel cells, etc.) are still in the development stage. In the next two or three decades these methods cannot be expected to play a role of importance in practical power generation.

Thus it will be accepted in this discussion of cooling water problems, that electric power in the next two or three decades will have to be generated mainly in the conventional way with steam turbines with their inseparably connected cooling water needs.

A modern fossil fueled power station discharges with the cooling water about 280 Mcal/s for a generated power of 1000 MW. Thus assuming, for instance, a temperature rise of 7° C, 40 m³/s of additional cooling water is needed. Present nuclear power stations discharge about 50% more heat with their cooling water, but in the future their efficiency might be expected to increase.

In the following a cooling water need of 50 m³/s will be assumed for 1000 MW of generated power, so a certain percentage of nuclear power stations is included. This percentage may rise with an increasing efficiency of these nuclear stations.

Alternate methods of cooling appear to be:

1. "Once through" fresh water cooling, in case of a sufficient supply of fresh cooling water from a lake or a large river which water may be discharged after heating in the condenser in such a way that it can never reach the intake again.

2. Once through salt water cooling. Taking in water from the sea or the ocean and discharging it at a suitable point may also give a simple once through cooling with some additional problems (corrosion, mussels) which do not occur using fresh water.

2. "Aeolic Power Generation in the U.S.S.R.," *Contacts Electriques*, No. 80, November 1969, p. 29.

This "once through" cooling only applies for one single power station. As soon as there are more power stations on the same river, lake or seacoast, one has to deal also with the next type of cooling:

3. Surface cooling. In surface cooling, water is taken from a lake and is discharged to the same lake at a remote point, such that it may be used again for cooling after it has had the opportunity to cool down by transfer of heat from the water to the atmosphere at the interface.

4. Cooling towers. These can be of the wet type, in which the heated water is cooled down in direct contact with the atmospheric air, mainly by evaporation; or of the dry type where the water is passed through large radiators in which the heat is transferred to the air without evaporation losses. In both types the air may have natural draught or forced draught.

5. Combinations of the aforementioned cooling methods.

Economic evaluation of the alternative cooling methods is very difficult as it depends on many factors that may vary widely from country to country and even from site to site in the same country.

Although not applicable to Europe, the results of an evaluation for America given by Hauser[3] will be mentioned here as the results are presented in a very illustrative manner in Table 3.

Table 3

Cooling type	Equivalent transmission distance in miles
fresh water	Base
cooling ponds	34.3
sea coast	13.2
wet cooling towers, mechanical draft	82.5
wet cooling towers, natural draft	78.2
dry cooling towers	321.6

These figures may depend largely on the circumstances, such as ground prices for cooling ponds, etc. However, the result that wet cooling tower costs correspond to an equivalent distance of about 80 miles and dry tower costs to about 300 miles gives a fair idea of the relative costs. With the same restrictions as for the evaluation of Hauser, it may be said that wet cooling towers cause an increase of 5–10 % and dry cooling towers of 10–15 % in the kWh price.

3. L. G. Hauser, *Cooling Water Sources for Power Generation*, Presented to the Southern Electric Exchange Production Section Meeting, Hot Springs, Arkansas, April 13–14, 1970.

The Cooling Water Situation in Western Europe

Most European countries estimate that their electricity requirements for the year 2000 will be several 100.000 MW, whereas the cooling capacity of most European rivers could better be expressed in units of 10.000 MW. It is thus clear that the rivers will be insufficient to meet the cooling water needs of the total power production in the future.

For countries such as Italy, Great Britain and Denmark, where most locations are no farther than 100 or 200 km from the seacoast it may be expected that in the future many power stations will be situated at the seacoast, unless other circumstances impede this development, as for instance in the Netherlands. For less favourably situated countries as France and Germany where much of the interior is more than 100 or 200 km from the coast, the cooling water reserves of the rivers as well as of the lakes must be exploited as intensively as possible. Further power stations will depend on cooling towers.

Although it seems unlikely before the end of this century, a possible change in the situation might be caused by the development of cryogenic cables. Such cables, whether superconducting or not, could make long distance electricity transport cheaper, so that the electric power in more instances might be generated in regions with enough cooling water.

After this short review of the general situation of the cooling water needs and reserves in Europe, the various cooling methods, their problems, the study of these problems and their importance for power production will be more thoroughly discussed for the Netherlands.

Once through Fresh Water Cooling on Rivers

At first sight the once through cooling water problems on a large river like the Rhine or Waal seem to be very simple. The cooling capacity of the river at a certain point can be easily calculated from the flow of the river Q m³/s and the allowed increase in temperature ($\triangle$ T° C) of the water.

For a Rhine flow of 1000 m³/s (mean value 2300 m³/s) a summer temperature of 24° C and a temperature of 30° C at the cooling water outlet, the cooling capacity of the Rhine would be Q.$\triangle$ T $= 6000$ Mcal/s. This cooling capacity would be sufficient for conventional power stations of more than 20.000 MW.

However, there are many problems involved in effectuating this cooling capacity. All these problems are studied in close cooperation by Rijkswaterstaat, (Government Service for Public Works and Water Control), RIZA, (Government Institute for Sewage Purification and Industrial Waste Treatment), the involved power stations and KEMA, (Joint Laboratories of the Electric Utilities in the Netherlands).

The building of one large power station of 20.000 MW on the Rhine would require that the total flow of the river be passed through the condensers. The many problems related to such a project, which could only be solved by building a dam across the Rhine with locks for shipping, will not be discussed here. For power stations of 2000 to 3000 MW about 100–150 m³/s of the river flow has to be used for condenser cooling. Model measurements by the "Waterloopkundig Laboratorium" (Hydraulics Laboratory) show that, even in the case where in- and outflow are very close to each other, recirculation will be no problem for such power stations on the Rhine or Waal. However, in other rivers with a lower velocity this recirculation may spoil the "once through" cooling, with part of the heated water passing again through the condensers.

The utmost care should be taken to prevent hindrance to shipping by cross currents near in- and outlet. The solution of this problem may depend on the local conditions, but is expected to be attainable in most cases.

There is a considerable sand transport over the bottom of the Rhine and Waal. Disturbances in the flow pattern may cause changes in sand transport leading to shoals in the river that are unacceptable for shipping. Studies on this very difficult problem are still in progress but seem to show that sand deposition will be a bottleneck for large power stations (3000 MW or more) on the Rhine and Waal. A possible solution might be the displacement of deposited sand by suction dredging.

As these problems demand urgent solution, simultaneous investigations are being made on the recirculation for a power station with its cooling water inlet downstream of the cooling water outlet. These studies are made both in a model and near the existing power station at Nijmegen. Although sand problems are expected to be much smaller and the recirculation is expected to be, from preliminary results, not more than 30% (for a cooling water flow of about 10% of the total river flow), this recirculation will certainly aggravate the problems mentioned below.

The mixing of the discharged cooling water with the river has been studied by means of temperature measurements near the power station at Nijmegen on the Waal and near the power station at Harculo on the IJssel. The mixing process proceeds very slowly: on the Waal the heated region only reached the centre line of the river at a distance of about 16 km below the point of discharge. The mixing process could roughly be described by turbulent mixing with a horizontal diffusion coefficient of the order of 0.2 m²/s. Figure 1 gives the temperature distribution across the Waal at a distance of about 12 km below the point of discharge from the power station at Nijmegen.

The investigations of the mixing process show that it is advisable to locate power stations alternatively on the right and on the left bank of the river.

One of the most severe problems facing the Netherlands in their future use of the Rhinewater for condenser cooling is the large number of power stations that Switzerland, Germany and France intend to build on the River. The problems arising from this common use of the Rhine for cooling purposes are discussed in a special working group on "Thermal Pollution" of the International Rhine Commission.

The consequences for the Netherlands will be that the incoming Rhine water will have a higher temperature than the natural maximum summer temperature of 24° C. The result is that the theoretical cooling capacity decreases and that greater quantities of cooling water are required notwithstanding limitations because of the sand transport problems already discussed. Thus an increase of the maximum summer temperature of the Rhine of only one or two degrees centigrade will cause severe difficulties for the use of the Rhinewater for cooling purposes in the Netherlands. The surface cooling (to be discussed hereafter) for a river like the Rhine gives only very slight relief for the Dutch power stations: at a flow of 1000 m³/s only ¼ of the temperature rise is cooled down over a distance of 100 km, a very long distance in a small country like the Netherlands!

Once Through Cooling with Seawater

Seawater cooling presents technical problems which can be solved at the expense of extra costs, but the essential difficulty is making it a real "once through" cooling. In the Netherlands there are only a few places where this is possible without prohibitive civil construction costs. One of these locations is near Rotterdam on the "Maasvlakte", where a larger power station will be constructed. Model studies have made it clear that recirculation of heated water to the inlet is negligible at this site. At other locations on the Dutch coast the sea is not accessible for cooling water purposes due to the dunes and the beaches. In addition enormous sand transport along the coast would make the constructions required to prevent recirculation very expensive.

The best method of using the North Sea for cooling purposes would be by locating power stations on the Wadden islands or on artificial islands. The latter might be a good solution in the far future if these artificial islands also have harbour facilities for large tankers and space for industry causing environmental problems. At the moment the costs would be prohibitive and the electricity transport from the islands, artificial or natural, would be beyond the present technical possibilities.

Another possibility of using seawater cooling, however not as "once through" cooling, lies in the estuaries of the Scheldt and the Ems. Intensive tide movements in these estuaries cause the discharged water to pass along the inlet making some recirculation unavoidable. However, by careful siting of the inlet and the outlet, mixing with the tide waters may be so intense that

51

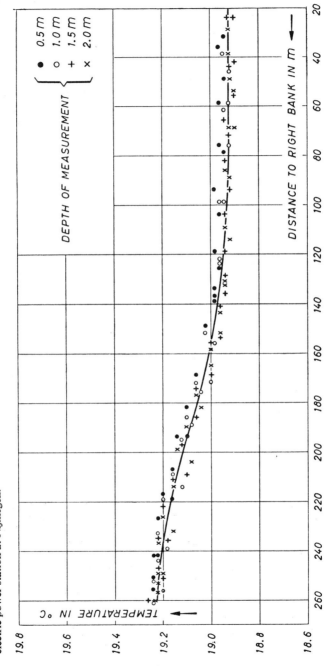

Figure 1. Temperature distribution across the river Waal, about 12 km below the cooling water discharge point of the electric power station at Nijmegen.

the increase of the inlet temperature is acceptably small. Moreover stratification may be used to avoid recirculation. On both estuaries the construction of large power stations has been started.

Surface Cooling

The potential use of stagnant water for surface cooling depends on the heat exchange processes at the water-air interface. These processes are many and they vary with the time of the day and throughout the year. Many investigators have tried to calculate the heat exchange coefficient due to these processes,[4] but their conclusions show considerable differences.

Direct measurements of the heat exchange coefficient in a cooling water circuit would be necessary to decide between the various theoretical results. However, for a number of reasons direct measurement is very difficult.

Several experiments have been made by KEMA (Joint Laboratories of the Electric Utilities in the Netherlands) on this subject and the first results of the most promising experiments are expected in the near future. Preliminary results show no evidence that the data will differ much from the value of 10^{-5} Mcal/m²·s·°C for the heat exchange coefficient determined in a very elegant manner by Wemelsfelder,[5] using routine temperature measurements.

Cooling Towers

Where natural cooling capacity on surface waters is insufficient, cooling towers can be used, although with higher electricity costs.[6] The use of cooling towers in the Netherlands has, however, a number of disadvantages that follow from the fact that the country has very high densities of: population, roadnetworks, built-up areas, waterways and harbours, electricity consumption,[7] industries and nature reserves. These facts make the use of

4. A. Becker, "Die thermische Belastbarkeit der Binnengewässer und die bei Durchflusskühling eintretenden Nutzungsverluste," *Besondere Mitteilungen zum Gewässerkundlichen Jahrbuch der DDR*, No. 2, 1965; J. M. Raphael, "Prediction of Temperature in Rivers and Reservoirs," *Journal of the Power Division of the American Society of Civil Engineers*, Vol. 88, PO2 (1962), 157; and D. K. Brady, W. L. Graves, Jr., and J. C. Geyer, *Surface Heat Exchange at Power Plant Cooling Lakes*, Cooling Water Studies for Edison Electric Institute, Report No. 5., November, 1969.

5. P. J. Wemelsfelder, "Wordt warmtelozing door centrales in de toekomst een probleem?" *De Ingenieur*, Vol. 80, No. 51 (1968), B179.

6. Bakker and Went, op.cit.

7. H. R. A. Wessels and J. A. Wisse, "A Method for Calculating the Size of Cooling Tower Plumes," *Atmospheric Environment*, 5 (1971), p. 743.

cooling towers unattractive for the Netherlands and their use on a large scale nearly impossible.

Cooling towers are large buildings which will be difficult to accept in a flat countryside.

One of the most serious problems for the Netherlands might be the long cooling tower plumes which may occur under certain meteorological conditions. This problem has been studied theoretically by the KNMI (Royal Netherlands' Meteorological Institute) who estimate that in certain circumstances the plume length could be several kilometers. The theoretical results should be verified experimentally, but it is clear that these plumes are undesirable in a country with a high density of built-up areas and of traffic on the roads and waterways.

Dry cooling towers produce no plumes but the other problems are more severe due to larger size and considerably higher costs.

The difficulties attendant on the use of cooling towers in the Netherlands make it important to use the natural cooling reserves as far as possible. In special cases however, where the natural cooling capacity is not sufficient, some additional cooling towers may be used.

Beneficial Thermal Effects

There is presently an inclination to mention only the more or less detrimental effects of cooling water discharges. In this section some attention will be paid to possible beneficial effects.

Thermal discharges may keep waters ice-free during at least a part of the winter. This offers advantages for shipping and runoff of rivers and is biologically important for water-birds and, because the aeration of the waters is impossible under an ice-cover, it is also important for fish.

The increase of water temperature by thermal discharges may prolong the period during which phenols are decomposed.

Even the accelerated decomposition of all organic wastes may be an advantage as long as the oxygen content is sufficient, but too much waste and temperatures that are too high lead to serious detrimental effects due to oxygen deficiency. In some cases it may be advantageous to aerate the discharged cooling water to improve the oxygen content.

The cooling water discharges may be used in fish farms and for the accelerated growth of shrimps and lobsters. The use of cooling water for agricultural applications has also been considered.

The combination of electricity generation with water desalting by flash evaporation causes more discharged heat than the generation of electricity alone. It is true that fresh water is gained, but the thermal problems are increased. The use of the discharged heat of electric power stations for evaporation and thus for concentration of sewage has been suggested. By these methods two problems might be solved at the same time, but it is to be feared that the technical problems for this procedure will be so difficult that evaporation of sewage water by direct heating with gas or electricity would be much easier. The same might be true for the sometimes suggested use of cooling water for the heating of buildings or for irrigation of agricultural fields with warm water; the availability of and the demand for heat do not coincide, and the transportation of low-grade heated water is expensive.

AIR POLLUTION FROM COMBUSTION PRODUCTS

by *A. J. Elshout* and *H. van Duuren*

Summary

The combustion of fossil fuels in power plants produces emission of sulphur oxides, nitrogen oxides, ash particles, carbon, unburned (partially oxidized) organic materials and anorganic trace gases.
An evaluation of these air pollutants, their concentrations and their properties, is given.
Present and future emissions and their relationship to the fuel pattern are discussed.
Methods of controlling emissions are reviewed.
Attention is given to the control of ground level concentrations of air pollutants. In this connection stack height, plume rise and meteorological factors are discussed.
Finally research and development contributing to a better understanding and control of emission and dispersion of pollutants in the atmosphere are considered.

Combustion Products

Combustible materials containing carbon and hydrogen have for centuries furnished man with a versatile source of heat and convertible energy. In recent years he has, to a large extent, been weaned from conventional solid fuels to the more convenient liquid and gaseous hydrocarbons.
Although nuclear power is likely to occupy an increasingly prominent position in the coming decades hydrocarbons will certainly continue to provide a significant portion of power supply. When fossil fuels are burned

chemical oxidation occurs, the combustible elements of the fuel being converted into gaseous products, and the non-combustible elements into ash. The gases resulting from combustion essentially contain nitrogen, water vapour, carbon dioxide and oxygen—all of which are already present in the atmosphere—in addition to certain air contaminants.

The air contaminants produced by fuel combustion can be divided into five categories: sulphur oxides, nitrogen oxides, ash particles, unburned, partially oxidized, organic materials, and anorganic trace gases, such as hydrogen fluoride (HF) and hydrogen chloride (HCl).

The sulphur oxides are directly related to the composition of the fuel used. The sulphur content of coal and oil ranges from less than 1% to 5% and more by weight, depending on the origin of the fuel. Between 95% and 98% of this sulphur is converted into sulphur dioxide (SO_2) on combustion.

In every combustion process the high temperatures at the burner result in the fixation of some oxides, due to oxygen and nitrogen combining to form nitric oxide (NO): $N_2 + O_2 \rightleftarrows 2NO$

In boiler practice most of the nitrogen oxides are emitted as nitric oxyde, in quantities kinetically rather than thermodynamically controlled. This NO can be oxidized in the atmosphere as a result of the reaction:

$2\,NO + O_2 \rightarrow 2\,NO_2$

Nitrogen dioxyde (NO_2) is the most toxic of the oxides of nitrogen and is an important component of the chemicals in photochemical smog. For these reasons it is usually assumed that all the emitted nitrogen oxides (NO_x) consist of NO_2.

In contrast to the emission of SO_2, where the amount released from any given equipment increases proportionally to the amount of fuel fired, the nitrogen oxide emission shows a more than proportional increase as production expands. One of the frustrating things about these emission rates is that they result from equipment designed and operated in accordance with the most efficient and up-to-date boiler practices. Some newer gas-fired power generating units in the Netherlands showed rather high emission values, which should be attributed to the development of higher combustion efficiency. This also explains differing emissions from combustion units of different sizes. For example, the median values given for nitrogen oxide emission by oil-fired combustion units used in power stations, industrial and domestic heating units are 1000, 750 and 125 g/Gcal respectively.

Particulate emissions from coal-fired units consist primarily of carbon, silica, alumina and iron oxide in the fly ash. The ashforming mineral matter in coal can range between 4% and 20% (by weight).

The shape of the particles often gives an indication of the type of firing unit used, along with its combustion efficiency. Pulverized coal units generally produce small particles, which are glassy and spherical while other types of stokers and domestic units produce larger particles, which are flakier and agglomerated.

Particles emitted from fuel oil combustion consist in general of 10–30% ash, 17–25% sulphates, and 25–50% carbon "cenospheres", formed during combustion. The particle size distribution is variable, ranging from less than 1 micron to 40 microns, but for the most part remains under 5 microns.

Hydrocarbon combustion is brought about by two different processes: hydroxilation and decomposition. Both processes occur during combustion in large power stations. Smoke from an oil-burning unit, for example, is the result of incomplete combustion. The chimney plume will then contain carbon, carbon monoxide, aldehyde and hydrocarbons. Large power plants are usually efficient in operation when controlled and regulated to achieve complete combustion; the emission of unburned or partially burned hydrocarbons is then not significant.

The source distribution of air pollutants is at the moment not well defined in European countries. For Western Europe we can use approximately the same values as those given for the United States by the U.S. NationalAcademy of Sciences. Table 4 gives this source distribution in percentages for the United States and also the proportion attributable to power station emissions.

Influence of the Fuel on Pollutant Emission

The combustion of coal and oil in power plants, in addition to releasing nitrogen oxides, fly ash and soot, is a major source of sulphur dioxide emissions. The combustion of natural gas produces only an emission of nitrogen oxides. The total emissions of these various components are primarily determined by the fuel balance.

Table 5 shows the consumption of fuels, for the years 1967–1969 inclusive, for the production of electricity by conventional thermal power stations (both public and industrial) in the EEC. Within this short space of time there was a clear tendency towards a decrease in coal consumption and an increase in the consumption of fuel oil and natural gas. This trend is expected to continue.

Table 4. Source Distribution in Percentages for Selected Air Pollutants in the United States[a]

Source	carbon monoxide	sulphur oxides	hydro-carbons	particulate matter	nitrogen oxides
transportation	91,7	2,1	64,7	15,0	38,7
industry	2,8	37,8	24,6	50,0	20,0
generation of electricity	0,8	44,5	0,7	20,0	30,0
space heating	2,7	14,8	3,3	10,0	10,0
refuse disposal	2,0	0,8	6,7	5,0	1,3
Total	100,0	100,0	100,0	100,0	100,0

a) W. C. Lepkowski, *Chemical Engineering News*, Vol. 46, No. 11 (1968), 8A.

Table 5. Consumption of fuels for the production of electricity by conventional thermal power stations in the Common Market countries (public and industrial).[a])

	1967	1968	1969
Total calorie equivalent in Tcal	811.000	870.000	974.000
Fuel in %:			
coal	66	63	58
oil	24	25	28
natural gas	4	6	8
industrial gas	5	5	5
waste	1	1	1
total	100	100	100

a) "Energy Statistics," *Statistical Office of the European Communities Quarterly* (1970), Nos. 1 and 2.

There is, in the Netherlands, a growing use of natural gas for electricity production, as Figure 2 shows. On the basis of rising electricity demand and forecasts of coal, oil and natural gas consumption in Dutch power stations, a calculation was made in 1969 of the future emissions of atmospheric pollutants over the next ten years.[8] Figure 3 gives the results of these calculations, which show that there will be no increase in emissions during the ten-year period. In fact, with the continuing increase in the use of natural gas, the total discharge of air pollutants will actually decrease. The curvature given in the figure for SO_2 is calculated for a sulphur content of the oil of 2,4% (median value of the oil used in the Netherlands in 1968). If the sulphur content decreased to 1% in 1980 the fall-off in SO_2 emission would become still more pronounced, as shown by the do ted line. That this does not apply to other countries can be seen from Figure 4, where the expected SO_2 emissions are given for the United States and Western Germany, as well as for the Netherlands.

8. A. J. Elshout and H. van Duuren, *Electrotechniek*, Vol. 47 (1969), 377.

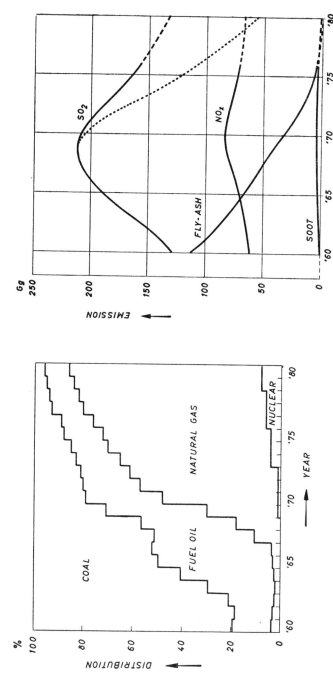

Figure 3. Total emission of air pollutants from Dutch power stations(1Gg = 10³t).

Figure 2. Distribution of fuel consumption for power generation in the Netherlands.

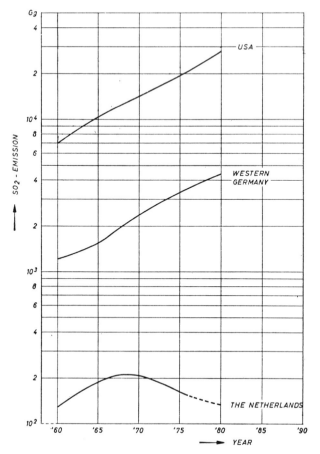

Figure 4. Total emission of sulphur dioxide from power stations in the United States, Western Germany, and the Netherlands ($1Gg = 10^3t$).

Methods of Controlling Power Plant Emission

The air pollution problems associated with electric utility plants burning fossil fuels result mainly from the emission of three pollutants: sulphur oxides, nitrogen oxides, and particulate matter. Technologies for the control of these pollutants are in various stages of development. The potential for improving these technologies depends upon the nature of the pollutant in question.

In the case of power stations the particulate matter to be removed is mainly fly ash. Various types of separators are used to remove the fly ash from the flue gases, and they can be grouped in three classes:

1. inertial separators — efficiency 70–85%
2. wet collection devices — efficiency 90–94%
3. electrical precipitators — efficiency 92–99,8%

The inertial separators used are cyclones, multicyclones, or a combination of both. All separators installed in the Netherlands since 1962 have efficiencies in excess of 92%. Particulate control is a well-developed technology, and all but the smallest of the submicron particles of fly-ash can, in principle, be removed by proper control of high-efficiency equipment.

There are several methods of reducing SO_2 emission; among them:
1. substitution of one fuel by another with a low sulphur content.
2. removal of sulphur from fuels before combustion.
3. removal of sulphur compounds (formed during combustion) from flue gases.
4. use of new combustion processes which themselves control the pollution.

Bearing in mind that the production of natural gas, low-sulphur and low sulphur coal can only meet a small part of the rising fuel demand in the world, we must next turn to the removal of sulphur compounds.

At first sight, the desulphurization of relatively small amounts of fuel would appear to be very much more effective compared with the removal of SO_2 from very large volumes of the gases. However, it has been established that, for the present, the removal of sulphur from fuel is both difficult and extremely costly.

In removing SO_2 from flue gases a distinction is made between wet and dry processes. A serious disadvantage of the wet processes is the low stack gas temperature, resulting in a low plume rise and, in some cases, local ground level concentrations of SO_2 which are even higher than those resulting from non-treated gases. This difficulty does not apply to the dry processes. However, the relatively low SO_2 concentrations in the flue gases, the large volume and high velocity of the gases, and the required reaction time necessitate very large installations. Throughout the world more than 60 processes have been proposed and are in various stages of development.

Preliminary economic evaluations of five of these processes, indicate that capital and gross operational costs will raise the total operational production costs by about 10%.

From kinetic and thermodynamic considerations it seems that the best method of reducing NO emissions is that of modifying the combustion processes.

In the case of two-stage combustion a reduction in NO_x emission of approximately 40% has been achieved.[9]

9. D. H. Barnhart and E. J. Diehl, *Journal of Air Pollution Control Association*, Vol. 10 (1960), 397.

Most of the modifications result in a measurable loss of combustion efficiency, with the exception of the twostage combustion method.

The Dispersion of Effluents from Tall Stacks

Elevated stacks, which disperse and dilute stack gases before they reach ground level, play an important part in air pollution control. They are effective in lowering the ground level concentration (imission) of pollutants, but do not in themselves reduce the amount of pollutants released into the atmosphere. Determining the expected maximum ground level concentrations at various distances from the plant site with various stack heights requires a sound knowledge of the meteorology of the area concerned. The behaviour of plumes emitted from stacks is primarily controlled by the stability of the atmosphere, which in its turn is governed by the diurnal cycle. In addition the vertical plume spread is related to the temperature profile along the stack.

Research and Development: Future Aspects

Control and improvement of environmental quality is of growing concern to the industrialized nations. One of the primary aims of the clean air policy has always been to minimize the ground level concentration of pollutants. For the main part the measures taken consisted in increasing chimney height. Generally speaking, in 1960 stacks were lower than 100 metres; now they rise 150 metres or more. However, little information is available which can be related to plume dispersion from the taller stacks. To remedy this deficiency a number of field studies are now under way in several countries.

Increasing pressure has been brought to bear on the limiting of emissions and has resulted in a sharpening of the control measures of the total emission. The possibilities of using substitute fuel to reduce the emissions of SO_2, for instance, are generally limited. The use of the vast reserves of high sulphur coal (particularly in the United States) and of high sulphur oil (particularly in Japan) in the utility market requires the development of a variety of SO_2 removal techniques. It is expected that, especially in those two countries, another 2 to 5 years will pass before one or more of the advanced sulphur dioxide control systems can be considered as proven technologies. They will also partly come into use in some European countries.

Research on, and development of the prevention of nitrogen oxide formation will be directed towards finding a method of reducing nitrogen oxide emission without a measurable loss in combustion efficiency.

One important factor in the clean air policy affecting the ultimate choice of possible measures is the matter of large-scale effects. The possibility that an increase in atmospheric pollutants would have a significant effect on the climate is under discussion. Thus there has been found to be an actual increase in the CO_2 content of the atmosphere, but the "greenhouse effect" seems to be small. The long-range effects of CO_2 are likely to be reduced, partly because of the buffering action of the ocean, partly because of the increased photosynthetic absorption and storage by forests, while a limit to CO_2 production is also set by the prospective exhaustion of fossil fuels. These discussions have not yet produced final conclusions, which emphasizes the need to obtain improved data. A definitive evaluation regarding the possible effects of the original and transformed combustion products is urgently needed.

BIOLOGICAL EFFECTS OF COOLING WATER DISCHARGE

by *J. L. Koolen*

Summary

The quality of the surface water must be dealt with carefully. At first a description of the concept of this water quality is given, defined by three parameters: oxygen, organic material, and fertilizer salts.
Then the influences on the water quality and the biological effects of cooling water discharges are described.
In the receiving water the biochemical and biological processes are increased at higher temperatures. Some examples are given. The biological consequences of the warming up of surface water may be serious, although in many cases no exact values are known at present.
Some additional, special effects due to certain circumstances are discussed such as the use of river water, of surface water, and of seawater.
Some recommendations are given for the planning of electricity production plants.
Careful deliberation between all people concerned is essential: the electricity producer, the water manager, the chemist, the biologist, the hydrologist.
Some general recommendations are: maximum permissible temperature (in the temperate areas) 30° C; control measures; quick mixing of the discharged coolant in the receiving water (hydraulics); no discharge of waste water in a cooling water system; avoidance of too large current velocities in the receiving water.

64

Introduction

Water is, in many of its reactions, a living substance. It is for this quality, due to the biological life present in it, that man uses it in many ways and in varying degrees of intensity. To guard against misuse is the water manager's task.

In this chapter the effects, in so far as they are known, of the discharge of large quantities of heated cooling water on the quality of the receiving surface water, will be examined. Also consideration will be given to technical control measures which may be required to prevent undesirable situations from arising.

The Concept of Water Quality

Surface water in its natural state, created by precipitation, rarely contains more than trace amounts other than H_2O. Additional matter is introduced by bottom leaching of substances, natural or man-made, and by the discharge of waste. Water, as such, has no "quality," which is an entirely relative concept used in comparing different waters or in contrasting one water before and after a particular purifying process. Water quality criteria are usually established by the content of oxygen, organic matter, nitrogen compounds, phosphates, various salts, various special organic substances etc. The most important criteria will first be dealt with in greater detail.

The *gases* present in the atmosphere, including oxygen, dissolve in water in certain proportions. Their solubility decreases as temperature increases. At a pressure of one atmosphere the solubility of oxygen in fresh water (the saturation value) is shown in Table 6.

Table 6

Temp. ° C	0	5	10	15	20
O_2 g/m^3	14.62	12.80	11.33	10.15	9.17
temp. ° C	25	30	35	40	
O_2 g/m^3	8.38	7.63	7.1	6.6	

The introduction of oxygen from the atmosphere into water is known as aeration or as reaeration. This is a spontaneously occurring, physical process. The rate at which it takes place is dependent upon the type of water flow (ranging from still pool to water fall), and upon temperature. It is directly proportional to the degree of undersaturation (the deficiency in respect of the saturation value). The degree of dependence of the reaeration on the type of water flow is illustrated in Table 7.

Table 7

Water type	Reaeration in grams $O_2/m^2/24$ hour period, per % undersaturation
small pool	0.01 – 0.02
large lake	0.04 – 0.05
slow flowing water	0.06 – 0.07
large river	0.09 – 0.10
rapid	0.1 – 0.2

These values are given for a temperature of 20° C. The rate of reaeration increases 1–2% with each additional ° C.

Apart from reaeration, oxygen is also introduced into water via carbon dioxyde assimilation. This process is done by phytoplankton, weeds and water plants. As the presence of these organisms is a result of the presence of fertilizer salts, the assimilation effects will be dealt with later.

Oxygen is an essential base material for a very wide variety of biological processes. The aquatic life*, built up in oxygenous water, ranging from micro-organisms to fishes, is dependent on the presence of oxygen. The degree of dependence for the various organisms is not clearly definable. However, a universally accepted criterion for the oxygen content is that fish (in Dutch waters, i.e. no Salmonidae) fare noticeably worse in water with an oxygen content of less than 5 g/m^3, although they can tolerate it for long periods. With a content lower than 3 g/m^3 harmful effects are increased to a marked degree, longer exposure greatly reducing the chances of survival.

Organic matter content of the water is affected via natural processes, such as the decay of organisms, and via the discharge of waste. Organic matter is used as food by a large number of micro-organisms, notably bacteria and fungi. Nature requires their help in converting waste organic matter into carbon dioxide and water (among other things), and thus to recirculate. This process is known as self-purification. In oxygenous water the bacterial flora growth implicates the consumption of dissolved oxygen to achieve this conversion process. With a natural supply of organic matter this consumption is rarely so large that it causes inconvenience to other organisms, such as fish. With a supply of organic matter from effluent discharge however, this relationship can be totally upset.

* The terms "aquatic life" and "organisms", used in the text of this chapter include the range mentioned here.

If there is no more oxygen present in the water, the conversion of organic matter continues, but via other groups of micro-organisms which are able to operate under the new conditions. In this way nitrate (NO_3^-) is converted into gaseous nitrogen (N_2), sulphate ($SO_4^=$) into the toxic and malodorous hydrogen sulphide (H_2S), and part of the organic matter itself into marsh gas (methane, CH_4). There is no doubt that these substances do not belong in healthy surface water.

If a certain amount of organic matter is present in the water this matter is converted by bacteria at a rate directly proportional to the concentration of this matter. The quantity which is decomposed per unit of time is thus greatest in the beginning and constantly decreasing. As soon as the water is undersaturated, the reaeration process begins, and the greater the undersaturation, the more intensive the reaeration. As time elapses the oxygen consumption due to the conversion of organic matter is decreasing so the oxygen increase from reaeration will eventually predominate. The result is that in water to which a certain amount of organic matter is added the oxygen content first decreases until it reaches a minimum value, and thereafter gradually rises, finally reaching the saturation value.

The bacterial decomposition of organic matter is so regular, at least in so far as any biological process can be said to be regular, that it can almost be considered as a chemical reaction. For this reason the organic matter content can be determined on the basis of the oxygen consumption of the water in a test bottle under control conditions. These conditions are: temperature 20° C, period 5 days, complete exclusion of light and air. The difference in the oxygen content before and after the test is called $B.O.D._5^{20}$ Biochemical Oxygen Demand at 20° C over 5 days.

The rate of the B.O.D. reaction increases approximately 5% for each additional °C.

The *fertilizer salts content* includes nitrogen compounds such as ammonium and nitrate, and phosphates. In water their effect is similar to that on land. They stimulate the growth of green plants: phytoplankton, weed and water plants, to a marked degree. This growth can become too exuberant if the concentrations of nitrogen compounds and phosphates are too high. Starting with only 1 gram of PO_4— and 6 grams of NH_4+ (or 20 grams of NO_3—) approximately 40 grams of vegetable matter can develop.

Mass growth of phytoplankton can seriously upset the delicate balance of aquatic life. Generally only a few plankton species will multiply excessively, crowding out the other species. Mass growth of higher plants (weeds, floating or rooted water plants, and riparian plants) can result in the water becoming completely overgrown, causing an acute deterioration in the living conditions of the other organisms.

Oxygen is formed by the growth of vegetable organisms, according to the well known photosynthetic reaction: CO_2 + H_2O + energy → organic matter + O_2.

This phenomenon is often considered a benefit for water quality. However, if organisms so formed die off later, this reaction is reversed. Then almost the same amount of oxygen is required as has been formed in the first reaction. Thus the overall effect of algae formation is nil.

So much oxygen is produced by excessive phytoplankton bloom that the water becomes oversaturated and gives off the gas to the atmosphere. As these organisms die off the water once again comes to contain so much organic matter, which is automatically cleared up by the self-purification process, that the oxygen content declines sharply and an equal amount of oxygen must again be captured from the atmosphere. It can even go so far that the water becomes totally deoxygenated. Moreover living phytoplankton also consume oxygen.

In the daytime this is not such a problem as they produce enough themselves. At night however, when this production ceases, their concentrated oxygen demand can have disastrous consequences for other oxygen-consuming organisms such as fish.

These same phenomena appear in the higher vegetable organisms. In addition, an abundance of floating plants, duck-weed for example, can just about entirely prevent reaeration and the entry of light into the water, causing serious difficulties for the normal aquatic life.

So one can say in general that a moderate presence of vegetable organisms (stimulated by slight amounts of phosphates and nitrogen compounds) is favourable, because of their oxygen production and their function as food for higher organisms. However, due to the presence of higher concentrations of fertilizer salts, an excessive amount of phytoplankton and higher plants can develop very quickly, making the water, as regards its oxygen content and the living conditions for other organisms, very unstable and disagreeable. Therefore, in general fertilizer salts must be considered harmful in their effects.

The Consequences of Discharging Heated Cooling Water

The *cooling water* used in a modern electric power station undergoes a temperature increase amounting to 6–8° C as a rule. This causes a decrease in the saturation value of the dissolved oxygen in the water. Saturated water will thus, by being heated, give off oxygen to the atmosphere. In the summer this can amount to: 9.17 g/m^3 at 20° C minus 8.07 g/m^3 at 27° C giving a release of 1.10 g/m^3. In winter this would be 13.84 g/m^3 at 2° C minus 11.59 g/m^3 at 9° C giving a release of 2.25 g/m^3.

When the oxygen content of the intake water is lower than the saturation value for the temperature of the discharged water, there will generally be no oxygen decrease. It is also possible to introduce special measures which supply oxygen to heavily undersaturated water in a power station.

Cooling water that passes through a power station is subjected to many mechanical processes. The water flow is turbulent, and it is reasonable to suppose that organisms carried along in the water are not all able to survive this treatment. Plankton in particular, with their delicate and graceful forms, would be easily damaged, and injury to unicellular creatures causes their death in many cases.

The result can be the formation of dead organic matter, which must later be cleared up by the bacteria in the water, causing oxygen consumption, which must therefore be considered as pollution. There are indications that these effects are less serious with larger generating units.

An aggravation of these effects takes its course in case an antifouling (= toxic) substance, usually chlorine, is either continuously or intermittently injected into the cooling water pipe system. The inevitable result is that a proportion of the living matter (plankton) flowing along in the water is killed, and thus converted into organic pollution. It is therefore good to promote the use of mechanical methods of antifouling (Taprogge-system).

All *chemical and biochemical processes* are accelerated by a factor of 1.5–3 per 10° C rise in temperature. This is the basis underlying all phenomena to be examined below. The acceleration continues with rising temperature in such a way that with higher temperature the processes become more unstable. More and more disruptions arise which can eventually kill off the organisms concerned. In this way the acceleration of the biochemical processes is overtaken at a certain moment by the lethal action of the high temperature. This is illustrated in Figure 5.

The temperatures for these phenomena vary widely for different organisms. But it is frequently found that temperatures in excess of 32–36° C will kill water organisms in the temperate zones in a relatively short time. Therefore a temperature of 30° C has been temporarily established as being the maximum permissible temperature for the cooling water to be discharged. There are indeed organisms known to be unable to endure even this temperature (for example fish species such as Salmonidae). But there are still no known data to indicate that this temperature would be unacceptable to the aquatic life in the waters mentioned.

Two points must be emphasized. First and foremost, it is not yet certain that this maximum value of 30° C will continue unchanged for ever. Research (begun in The Netherlands and other countries) will have to prove whether it is a permissible value. It is not impossible that this limit will have to be lowered in a few years time. Secondly, it would be unwise to expect that the value of 30° C will be significantly increased. If 32° C should be a lethal barrier, for example, a permitted temperature of 30° C is already very high.

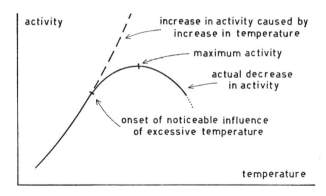

Figure 5. Relationship between biological or biochemical activity and temperature.

It is possible to illustrate this by comparing the effect of temperature with that of a toxic substance. If a substance in a concentration of, let us say, 1 g/m³ is lethal, that still does not mean that the water may be loaded with the substance to a value just below the lethal figure (0.8 g/m³ for example). For a number of reasons the water manager needs to observe wide safety margins.

The decrease in the saturation value of oxygen in the receiving water resulting from the discharge of cooling water is not so spectacular as that which takes place in the cooling water itself. The discharged cooling water is diluted with unused surface water, and the temperature naturally drops. If the discharged cooling water already has the saturation value corresponding to its temperature, then the question arises whether, and to what degree, the mixing of cooling and surface waters causes further oxygen loss. The curve of saturation value against temperature is, over a span of 8° C, practically linear. Comparison of water quantities, temperatures and oxygen contents show that further oxygen loss does not take place.

The rate of the B.O.D.-reaction increases approximately 5 % with each °C temperature rise. The rate of reaeration increases 1–2 % with each °C temperature rise. The oxygen undersaturation corresponding to a certain oxygen content decreases 1.5–3 % with each °C temperature rise. As a result the reaeration at different temperatures varies only little in numerical terms. The end result is that the decomposition of organic matter takes place more sharply, and somewhat earlier in time, at higher temperatures. Figure 6 illustrates this.

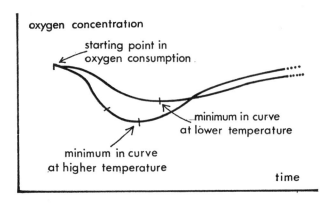

Figure 6. Influence of temperature on oxygen consumption curve.

When living conditions of oxygen-dependent organisms are concerned, the minimum oxygen content must be considered the limiting factor. For this reason a higher temperature must be considered relatively harmful. One advantage is that the pollutants are more quickly and completely cleared up.

Ammonium (NH_4^+) is a breakdown product of the conversion of organic wastes in a sewage purification plant or in surface water. It is also found in the aquatic world as a result of its use as an artificial fertilizer on the land. There are groups of bacteria which are able, with the aid of oxygen in the water, to oxidize ammonium to nitrate (NO_3^-), these are the so-called nitrifying bacteria. In this reaction, of course, oxygen is consumed from the water. This consumption is not included in that arising from the conversion of organic matter, the B.O.D. reaction. The oxidation of ammonium is dependent on the temperature; generally it takes place above 10° C, so it is a summer and not a winter phenomenon. The result of warming surface water by cooling water discharge is that on the one hand the period within which the resultant oxygen consumption occurs is extended, and that on the other hand the reaction will be accelerated, with a consequent increase in oxygen consumption per unit of time.

Since these results all contribute to bringing the oxygen condition of the water one step nearer to a critical level they must be considered as having a harmful effect on water quality.

The mud in most standing waters, particularly shallow canals and lakes, contains a great deal of organic matter which slowly decays. As oxygen from the water above does not generally penetrate very far into the mud layer, the decomposition of the organic matter in the mud occurs for the most part under oxygen-free conditions. In this way gases may be formed, among them methane and hydrogen sulphide. Increasing the temperature accelerates all reactions, also gas formation (several % with each °C temperature increase). The result of increased gas development in decaying mud layers is that the mud is stirred up by gas bubbles and brought into suspension, giving the water above a greatly increased amount of organic matter to dissimilate. This again makes demands on the oxygen.

In addition, the poisonous hydrogen sulphide comes into the water at the same time. These phenomena are met with on very hot summer days, often in shallow polluted waters, accompanied by widespread fish deaths. It is evident that such a situation is far from ideal, and it should be borne in mind that raising the average water temperature in polluted water increases the probability of its occurrence.

The production of phytoplankton from water, carbon dioxide, fertilizer salts and trace elements—that is to say from anorganic compounds—is stimulated by higher temperature, as is the growth of bacteria based on organic waste. These two types of organisms are in great demand as food for higher life forms. The increase in the basic food source initiates an intensification of the entire biota. If the food exceeds a certain level the intensification becomes so excessive that it can disturb the balance of water life. In addition the instability in the oxygen content mentioned earlier is encouraged by an over-exuberant flourishing of phytoplankton.

A rise of temperature accelerates the vital functions of aquatic organisms. Activity is increased in many respects, particularly in the assimilation of food and in the production of energy. This phenomenon can be put to very good use in fish farms, where the supply of heated water can solve many problems, particularly in winter. There is, of course, also a growing chance of organisms living "too intensively", with all the adherent disadvantages.

The propagation of many species of organisms is activated by temperature rise (springtime). So an artificial temperature rise will obviously have a disturbing effect, changing the normal behaviour of organisms for the time of year. Different organisms can react in different ways to a rise in temperature. If the various species under natural conditions are in harmony with each other, i.e. living *off* each other, these different changes may have an adverse effect on the aquatic life as a whole.

What has been described above may cause some species of organisms to die off, and others to appear in their place. Therefore changes in the population structure can be expected.

Like plankton, water and riparian plants can begin to display a faster and more exuberant growth as a result of increased water temperature. On the other hand the currents in the cooling circuit may inhibit such growth. The growth and blooming of riparian plants can move to earlier periods in the season, but in the long run it is quite possible that a sharp temperature rise could bring about a marked reduction of the flora.

Broadly speaking, the sensitivity of aquatic organisms to probably all harmful factors in the water increases with higher temperatures. There are negative effects to lethal or sub-lethal concentrations of toxic substances, to oxygen deficit, etc. Therefore, generally a temperature increase must be considered harmful to fish, for example, in addition to all the other adverse effects mentioned.

In *lakes* with standing water aquatic life, such as plankton, invertebrates and fish, is based on the absence of current. If, in such waters, a continuous current is introduced, as would happen if they are to form a part of a cooling circuit, then undoubtedly certain species will be unable to adapt, and they will disappear. In the long run they will be replaced by species which belong in flowing water.

Large lakes in The Netherlands are situated in flat areas, where the ground may have a loose composition of peat, clay, or sand. The introduction of a current here can have disastrous effects on water banks. A current with a velocity of more than 0.3–0.4 m/s is likely to stir up bottom mud containing a great deal of organic matter. This can have much the same effect as the bubbling up of gas from decomposing bottom mud in shallow waters.

The muddying of previously "clear" lakewater, however, can have a positive effect in preventing the excessive flourishing of phytoplankton. A great deal of sunlight is required for this process, and an abundance of mud particles can restrict the penetration of light.

For the rest, the transition from clear to muddy water can definitely bring about important changes in the population structure of aquatic life.

Discharged sewage, which can bring about poor water quality in standing water, is relatively well-diluted in running water. The adverse effects are thus spread out over a wider area and are thereby reduced in absolute terms (the oxygen decrease, for example, is slighter). From this point of view, therefore, a current in the water can be beneficial.

An exception to this is the contamination by sewage bacteria. These also are more highly diluted in running water than in standing water, but this dilution must be very high indeed before the bacteria can be disregarded. This contamination can result in a far greater body of water being made unsuitable for such activities as swimming than previously was the case.

One unarguably beneficial result of increasing current speed in a surface water is the increase in reaeration. Values for this have already been given in table 7.

Only *large rivers* can be thought of as a coolant source for modern electric power stations which use scores of cubic metres of water per second. Such a river will almost never be used exclusively for cooling purposes, but will have a wide variety of other uses at the same time. These include its use for drinking water preparation, for irrigation of agricultural land, as a way for the run-off of surplus rainwater, as a waterway for shipping, and as receiving water for wastes—purified or otherwise. The latter use in particular can cause a deterioration in water quality. To this much attention must be paid. Usually this means a fairly thorough purification of discharged sewage. However, even then, large rivers in most places receive the residual pollution (about 10% of the original organic matter). This cannot be directly eliminated by the purification plants, and must be further dealt with by the self-purifying processes in the river water itself.

In addition, the river becomes contaminated by fertilizer salts, and discharges of difficult or unpurifiable wastes from (chemical) industries. In the large population centres these wastes can cause such pollution of the river that the water quality for hundreds of miles can give the water manager a great deal of trouble. An effort must be made by more extensive purification and other control and/or technical measures to bring the water quality back to an acceptable state. On the whole, the aquatic life in such rivers is disturbed, unstable and strained.

The water quality in polluted large rivers, measured by the oxygen content, is definitely worse in summer than in winter. This is caused on the one hand by the lower river run-off in summer, and on the other hand by the accelerated and increased decomposition of all sorts of waste matter caused by the higher temperatures, which results in even more oxygen being withdrawn from the water.

Siting many large power stations on such a river, resulting in considerably higher average temperatures both in summer and winter, will reinforce the adverse summer situation as regards water quality, and extend the periods in which this occurs. The instability of the water quality will, of course, increase still further. For this reason much thought must be given to the planning of electric power stations. Associated studies will have to be carried out to ascertain the correlation between the increased electricity production, the increased river temperature and the water quality.

Should a large river continually have to serve a multitude of uses, as outlined above, then a careful assessment must be made as to whether the various forms of use, including use as cooling water, are not mutually exclusive or give rise to unacceptable conditions of water quality. The larger the part of the river run-off required to pass through a power station as cooling water (rain rivers in dry periods!), the more carefully the above assessment must be made. For example, chlorination would have to be discontinued in favour of alternative, mechanical antifouling methods.

Most *seas* are highly stable water masses which grow slowly warmer in the spring and grow slowly cooler in the autumn. Their chemical composition is, as a rule, perfectly constant. The stability of the sea milieu is in many ways much greater than that of fresh water, extreme situations being more rarely encountered. This is a natural phenomenon and the seawater life is adapted to it. The general impression is that seawater organisms are less able to withstand extreme conditions than are freshwater organisms, this applying particularly to abnormal temperatures or temperature fluctuations.

As the aquatic life in the sea must not be too much disturbed by the discharge of heated cooling water, the latter should be discharged in such a way that the heat is as quickly dispersed and released as possible. It is generally the case that the areas and quantities of seawater available are sufficiently large to enable this to take place easily. But the attendant civil engineering problems are not so simply dealt with.

Chlorination is the most commonly used method to rigourously combat the spontaneous growth of sea-weeds, sea-acorns, mussels and other testaceans in order to prevent impairments of pumps, pipes, channels, etc. for cooling water. The discharged cooling water will contain a certain residual amount of chlorine that will affect the organisms in the receiving water. An effort should be made to use as low concentrations of chlorine as possible and in addition, to disperse the discharged cooling water effectively in the receiving sea, spreading the effects out over a large area and thereby reducing their harmfulness in terms of absolute size.

A more or less physical phenomenon that also has biological aspects, is the salt-stratification of seawater along a coast where large rivers discharge. The cooling of discharged water takes place on the water surface. It is therefore necessary to carefully choose the intake water so the specific gravity of the discharged heated water is small enough to enable it to remain on the surface. Should it travel as an intermediate layer in the seawater, the heat will be, as a rule, dispersed more slowly. This may be avoided by careful site planning of intake and outfall points.

Recommendations

First and foremost it must be said that the present system of electricity generation, in which more energy in the form of waste heat is released to the environment (primarily the water environment) than is generated as electricity, can hardly be called ideal. There needs to be unremitting effort towards improvement, especially as regards the efficiency of electricity production and re-use of heat, in order to keep the production of waste heat at a minimum.

Some emphasis should be placed on the fact that many of the influences described in the foregoing sections are not yet quantifiable. It is only possible to show tendencies and orders of magnitude of the changes brought about in the water by waste heat. And where calculations are, or in the future will be possible, the necessary reservations must still be made as to the exactness of the results.

For this reason the water manager should be cautious about admitting relatively large quantities of hot cooling water into surface water under his care. This demands wide-ranging preliminary deliberation, study, and attention to recent developments, all in order to maintain the surface water at optimum suitability for its uses.

A few general recommendations follow. They are based on the premise that the situation occurring in the surface water concerned must be disturbed as little as possible. Notice that they specifically regard waters in the temperate zones.

A. Maximum permissible temperature.

The temperature of the water discharged by the power station should under no circumstances exceed 30°C. The temperature at various important points (certainly at the station inlet and outlet), should be continuously monitored. Should the temperature at a particular point threaten to exceed the permissible level, measures to solve the problem should be taken immediately.

This can be done by reducing electricity production or by using a larger amount of cooling water (not by mixing the overheated discharged water with cooler water!) The quality of the receiving water should be regularly inspected.

In the ad hoc Study Group on Water Conservation of the Council of Europe there is a paper on cooling water problems, under consideration at the moment, in which there is a proposal to limit surface water temperature to a maximum of 25°C. This demand, originating from Switzerland, is often impractical for waters in other countries. However, for waters upon which high demands are made (direct extraction for the preparation of drinking water, recreation) the recommendation has to be emphasized.

B. Mixing of discharged cooling water with the receiving water.

The discharged cooling water should be mixed as soon as possible with a relatively large quantity of receiving water. In this way temperature is reduced without waste of time, giving the indigenous water flora and fauna the best possible chance of holding their own. The hydrographic and hydraulic conditions around the discharge point therefore deserve close attention.

C. Discharge of sewage.

The discharge of sewage into a cooling circuit, and also into waters in open communication with it, should be avoided at all costs.

D. Current velocity.

The current velocity in a shallow cooling circuit should not exceed 0.1–0.2 m/s, in order to avoid stirring up bottom mud. Checks should be made that the banks of such watercourses are equal to this speed.

Should these recommendations give rise to measures that appear expensive, it may be said that it is a matter of the quality of life and no risks should be taken that would be avoided in a matter of life itself.

In the Netherlands a close contact exists between the electricity producers, the managers of the principal waters considered for the cooling of large generating stations, the State, and Provincial Authorities. These contacts are scientifically supported by a working group of representatives from the electricity producers, the State, and chemists and biologists from various institutes in the Netherlands engaged in surface water research. Particularly because of the international contacts on cooling water problems which have been recently developed (for example, concerning the use of one river by several countries), it is hoped that a thorough handling of these problems may become standard international procedure.

BIOLOGICAL EFFECTS OF AIR POLLUTION

by *P. E. Joosting* and *J. G. ten Houten*

Summary

Exposure of living organisms to sulphur dioxide, sulphuric acid, fly ash, other particulates, and oxides of nitrogen is discussed from the points of view of air pollution phenomenology, specific and nonspecific responses of plants, animals and man, and environmental and constitutional factors that influence the mode of response and the grade of effect. Some examples of dose/effect relationships are given. For sulphur dioxide and (black) suspended matter (as indices of pollution from the use of fossil fuels) both tolerable and unacceptable criteria are tentatively suggested for the purpose of planning and control.

In addition argument is given—from the biological, physiological and common sense standpoints—in support of the concept of clean air conservation.

Introduction

The effect of atmospheric pollutants upon living material is determined by a number of factors:

physical and chemical properties of the pollutant;
its concentration;
duration of exposure;
environmental conditions;
susceptibility of the organism;
locus and mode of uptake;
metabolism and rate of elimination.

The latter is of particular interest in the case of human beings and many animal species: a great number of substances may pass through the body or can be eliminated, in their original form or as a detoxified metabolite. Vegetal life usually has no such facilities at its disposal, although a type of elimination of non-physiological substances may take place, e.g. by transporting sulphates into the roots after the leaves have been gassed with sulphur dioxide.

In view of the biochemical rules and probabilities that determine life's fundamental processes, it should be recognized that most pollutants can be characterized as "unknown" and strange to the natural system, either as substances per se, or in the concentrations encountered. This means that life in modern environments has to cope with a great number of chemicals that are irrelevant from the point of view of nutrition and energy demand.

Therefore those substances can be considered toxic or potentially noxious, and to get rid of them requires biological energy on the part of human, animal, and plant life. Detoxification and elimination processes happen to take place at the expense of an essentially limited potential of life's phenomena. Consequently pollution problems should be appreciated in terms of body burden and loading capacity, which can be assessed partially on the basis of objective, or of subjective and arbitrary criteria.

The toxicity of a substance can only be assessed if duration and pattern of exposure are known. Therefore a thorough knowledge of air pollution phenomenology is of fundamental importance. Of a number of pollutants, such as sulphur dioxide (SO_2) and suspended particulates, it is known that their ground level concentrations per measuring station fluctuate according to a pattern that is reproducible over longer periods, namely an approximately normal (= Gauss) distribution of the logarithms of 24-hour averages over a year.

In agglomerations with a spread of a great number of different sources of varying output, the geometric standard deviation of such a distribution of SO_2 data, for example, is in the order of 1.5. In the case of a measuring station near to a single source—such as the stack of an isolated power station—this figure will be in the order of five. In practice the net result may be that the geometric mean of both distributions differs to an extent of a factor ten, although on the other hand the probability of excessively high concentrations over a few days per year may be of the same order in both instances.

Although the response of living material is theoretically dependent on the integral of a pattern of exposures, the exposure can in practice be fairly accurately characterized by the mean concentration and the number and level of peak concentrations of short duration. Although no exact relationship exists between extremes of exposures, for practical purposes one may assume that, within a definite range of exposures and with not too toxic substances, the degree of effect will be more or less consistent, depending on the constancy of the sum of log concentration and log exposure time:
$E \sim k (\log C + \log T)$.

Substances that are capable of causing damage to living material and that originate from the burning of fossil fuel in power plants are SO_2, nitrogen oxides, fly ash and other suspended particles. A relatively small amount of sulphuric acid (H_2SO_4) may be emitted. A part of the SO_2 emitted into the atmosphere will also be converted into H_2SO_4.

Although the latter is precipitated fairly rapidly by growth of the hygroscopic droplets, these sulphuric acid aerosols should not be neglected as an important factor in the causation of damage to living and non-living material.

Effects on Plants and Vegetation

The effect of pollutants upon living organisms depends on several factors. In the case of plants an important role is played by the great differences in susceptibility existing between various species of both cultivated and wild plants, and even between varieties or individual specimens of some species, in exceptional cases. Moreover, the influence of environmental conditions such as climate, radiation or soil chemistry and moisture, is far more pronounced than in the case of man and the higher animals. These influences vary, however, according to the types of pollutants involved. In this respect the scope of the present survey is rather limited.

In cases of coal burning, especially, particulate matter contributes to undesirable deposits on vegetables and fruits. Although direct injury is seldom caused, the harvested product must be cleaned before marketing.

This has financial consequences for the grower. In areas with greenhouses and dutch lights, solid deposits may considerably reduce the amount of light passing through the glass coverage, thus diminishing growth and crop yields.

However, gaseous pollutants are of greater importance—particularly sulphur dioxide and, to a lesser extent, nitrogen oxides. Some countries like the Netherlands are in a fortunate situation from the agricultural point of view: natural gas, which is free of sulphur compounds, is available for heating purposes and electric energy production. As long as this short-term source of energy is not exhausted it offers an alternative to coal or mineral oil as a fuel in areas where current sulphur dioxide concentrations might be particularly unacceptable.

Careful investigations of the influence of various environmental factors on the susceptibility of plants to *sulphur dioxide* have resulted in some firm conclusions. Meteorological and soil conditions play an important role as to the degree and amount of damage caused by certain concentrations of SO_2 present in the air over known periods of exposure. A high humidity increases the susceptibility of plants: from 30 to 50% relative humidity (R.H.) there is a gradual increase, and between 60 and 90% R.H. a rapid increase. The influence of temperature is less pronounced.

The amount of sunlight has no clear influence upon the effect of SO_2. This is in contrast to what has been observed in the case of plant damage due to peroxy-acetylnitrate (PAN), which is the product of a chain of photochemical reactions and an important constituent of atmospheres with an oxidizing (Los Angeles) type of pollution. So far as SO_2 under experimental conditions is concerned, the plants of lucerne (alfalfa) appear to be most susceptible about four hours after sunrise, on both sunny and cloudy days. With spinach and radish the same phenomenon has been observed. Experiments concerning the influence of soil moisture on the effect of fumigation with SO_2 have shown that vine plants, for example, can be badly damaged if the soil is constantly kept moist, whereas, even after several fumigations with high concentrations (4.5–9 mg/m³ or 1.6–3.2 ppm) of SO_2, no damage has been observed to plants grown on a dry soil. Similar results have been obtained with tomato plants. In colza the resistance to SO_2 increases with the mineral nitrogen content of the soil.

Acute symptoms of SO_2 injury are clearly visible in the interveinal leaf tissues which collapse and later become desiccated and bleached, while the veins themselves remain green. With small doses of SO_2 no symptoms occur as a rule, because SO_2 is metabolized in the leaves into sulphates which are at least partly transported towards the roots. If small amounts of SO_2 are absorbed over longer periods, the leaves may become chlorotic. Epiphytic lichens and mosses (bryophytes) are much more susceptible than higher plants. At an annual average concentration of SO_2 above $45\mu g/m^3$ ($\sim$ 0.016 ppm), only some species, such as Parmelia saxatilis, may survive. Therefore lichens may be used as indicators for the presence of SO_2. As a concentration of $100\mu g/m^3$, may, in the long term, adversely affect pine trees, landscape designers should be advised to be wary about planting conifers in a Parmelia saxatilis "desert".

Sulphur dioxide can act as a synergist in the causation of damage by ozone. This has been clearly demonstrated in American fumigation experiments using low concentrations of either pollutant, individually as well as in combination.

As a rule, the *nitrogen oxides* which are formed in the combustion process of conventional power plants will consist, in part, of NO_2 at the point of maximum ground level concentration. In short term experiments this gas may cause the same symptoms as SO_2 if high concentrations (about 5.4 mg/m^3 or 3 ppm) are used. In the outdoor atmosphere such concentrations of NO_2 rarely occur under normal conditions. In the Netherlands the former have only been found after, say, an accident at a chemical factory. At low concentrations (of about 0.2–0.5 mg/m^3 or 0.1–0.3 ppm) visible effects on tomato plants only appear after a continuous fumigation for several months. However, growth and production may be seriously affected, and can lead to a decrease in yield as high as 22%.

In some densely populated areas of the USA a concentration of 0.5 mg/m^3 (or 0.3 ppm) NO_2 is registered quite regularly. In Western Europe the NO_2 levels are much lower, and it seems improbable that the emissions from electric power plants will contribute substantially to the buildup of ambient NO_2 levels that may be hazardous or adverse to plant life. Only if the atmospheric conditions are favourable for the formation of photochemical oxidants like ozone, does NO_2 act as a "catalyzer", contributing to the ozoneforming process.

Effects on Man and Animals

As mentioned already, man and his domestic animals have a reasonable capacity to metabolize substances that are "strangers" to the physiology of their systems. Thus they can cope with a stream of biologically irrelevant information. But there are limits, as has so clearly been demonstrated during so-called smog disasters—periods of a few days with adverse meteorological conditions causing cumulation of pollutants (for example, during the heating season). During such periods (and sometimes shortly afterwards), a lesser or greater number of particularly vulnerable people appear to be afflicted, as shown by increased mortality, sickness rates and functional impairment (cf. Greater London 1952 with 4000 excess deaths in a few days, at an expected death rate of about 300 per day during wintertime). As to the causation of the nonspecific health effects in cumulation periods with a reducing type of pollution, the following substances are suspected to play a role: sulphur dioxide, black suspended matter (soot), and other particulates and aerosols including sulphuric acid and sulphates. Although not routinely measured, carbon monoxide (from open fires, stoves with stagnating chimney flow, traffic) could also have played a part during these periods.

On the one hand it is to be expected that under such circumstances sufferers from chronic bronchitis and those with impairments of the cardiorespiratory system would be the main potential victims. On the other hand it is surprising just how disastrous the effects have in fact been, when it is considered that they resulted from exposure to polluted atmospheres which in practice could only be characterized by such concentrations of SO_2 and particulates as have failed to demonstrate any unfavourable effect under comparable circumstances, like industrial exposures and experiments with human volunteers and animals. One should recognize the fact that SO_2, H_2SO_4, NO_2, CO, and particulates are the only measured indicators of an amalgam of pollutants that is, in effect, more toxic than can be deduced from the presence of only one or a few parameters.

Various substances may interact, or even separately potentiate their respective responses. In this respect it is very well known that small particles with a diameter of less than five microns can penetrate deeply into the airways and lungs, where they may irritate and evoke responses of an obstructive nature, depending on the acid gases or liquids adhering to the particle surface. Mucus production and reflex constriction of the smaller bronchi and bronchioli, together with the increased airway resistance and impaired breathing in respiratorily crippled and sensitive people, is a typical result of exposure to these essentially non-specific stimuli.

One should not be surprised if future investigations bring fresh facts to light that lead to new concepts in this field, as is the case with the latest ideas on the uptake and metabolism of SO_2 in the body. A few years ago SO_2 was only appreciated as an external agent that could irritate the mucous membranes of the airways when inhaled.

Modern investigations with tracer techniques and balance studies of uptake and output have shown that SO_2, being an easily soluble gas, is 90 to 100% absorbed in the aqueous surface of the mucous membranes of the nose and throat (nasopharynx). From there it is transported through the tissues into the blood stream. In various organs an accumulation of labeled sulphur molecules can be measured. After some time SO_2 appears in the expired air and is thus excreted, in a proportion not yet quantified.

A great deal of the energy transport in the body is performed by (de-) hydration of sulphhydryl (HS-) groups. One may assume that inhaled and circulating SO_2 influences this system that normally incorporates compensatory mechanisms.

A loss of vital energy may be camouflaged in healthy organisms, where in ill or disabled people it would become readily apparent.

Considerable attention has been paid to the relationship between mean and peak concentrations of SO_2 and (black) suspended matter on the one hand, and the mortality pattern in the population on the other. This also has been done for other health paramaters such as hospitalization, absenteeism and functional disturbances.

As so the causation or promotion of adverse health conditions due to atmospheric pollutants, it is probably unrealistic to think in terms of specificity and direct cause and effect relationships.

It should be mentioned in passing that it is an established fact that inhalation of cigarette smoke is the main contributor to the causation, promotion and continuation of diseases such as chronic bronchitis and lung carcinoma. Compared with the detrimental effect of cigarette smoking, the influence of air pollutants from combustion processes in their normal emission concentrations, can probably claim only 10 % of the joint influence of both these features of modern civilization!

Sulphur dioxide and other gases can be registered by the olfactory and central nervous system at concentrations sometimes far below the sensory threshold. The electroencephalogram (= record of the electrical activity of the cerebral cortex) appears to reflect subsensory registrations of changes in the composition of the atmosphere. With increasing concentration of the test gas an initial rise in the cortical activity can be observed, followed by a decline.

Although these findings look spectacular, it has not yet been evaluated precisely what practical significance they have as regards the possible inhibition of brain function. But one should recognize the fact that human individuals, as well as the community as a whole, have increasingly to rely upon the integrity of their sensory and intellectual information processing system.

The threshold for smell detection of SO_2 lies at a concentration of about 3 mg/m³ ($\sim$ 1 ppm). Unpleasant irritation may occur at a level of 5–6 mg/m³ ($\sim$ 2 ppm). Nevertheless, the generally accepted threshold limit value (TLV) for pure SO_2 in industrial working conditions is 13 mg/m³ (5 ppm). One should note that a TLV refers to time-weighted concentrations for a 7 or 8-hour workday and a 40-hour workweek. Because of wide variation in individual susceptibility however, a small percentage of workers may experience discomfort at concentrations at or below the TLV. An increase in airway resistance has been observed.

Although the working mechanism of SO_2 is still under study, it is possible that reflex responses may play a role in the constriction of the smaller bronchi. On the other hand, irritation of the mucosa does induce an increased mucus production which is an especially deleterious effect in bronchitics, who are liable to react on exposure to concentrations much lower than the TLV, e.g. in the order of 3 to 6 mg/m^3 (1 to 2 ppm), of SO_2.

One should be aware of interactions between SO_2 and various other substances before attributing a recorded harmful effect to the influence of a single substance like SO_2. In practice the presence of soot and particulates seems to play an important role. It has been possible to record unwanted respiratory symptoms in bronchitic patients who were incidentally exposed to an atmospheric pollution that could be characterized by a level of about 1 mg/m^3 ($\sim$ 0.4 ppm) SO_2 under presence of only 200μg soot per m^3.

Although such concentrations do not, in practice, occur regularly in areas with single sources equipped with an appropriate control technique, one should avoid considering SO_2 as being a substance with such a low toxicity that it can be neglected. Especially since it has become clear that SO_2 is absorbed completely and rapidly into bodily organs and systems, the question as to the vital significance of this omnipresent pollutant should be answered, and rapidly.

A *H_2SO_4 aerosol* is, in terms of S-equivalent, far more irritating and harmful to the respiratory tract than is SO_2. Exposure of bronchitic patients to H_2SO_4 concentrations of 250μg/m^3 may induce unpleasant irritation and shortness of breath. In healthy people a concentration of 120μg/m^3 over a limited exposure period would not give rise to clinical symptoms.

It should be mentioned that other aerosols, like sodium chloride or finely dispersed metal salts, also have the capacity to stimulate an SO_2 effect by the formation of a H_2SO_4 aerosol in the inspired air. This is of particular significance where a solitary source such as a power station emits SO_2 into a neighbouring area where considerable amounts of catalytically oxidizing substances are present in the atmosphere, as in the case of steel plants and metallurgical works.

Fly ash consists of a part of the incombustible minerals of the fuel, i.e. silicates and a great number of metal compounds. Other products, particularly of incomplete burning, may adhere to fly ash particles. For the time being fly ash as such is not a suspect material from the health point of view. It is possible that the metal compounds play a role as condensation nuclei and catalysts in the transformation of absorbed SO_2 into H_2SO_4. If these, extremely hygroscopic particles are small enough to be inhaled (diameter $<$ 5μm), fly ash may then play a role as a vehicle for the deposition of H_2SO_4 in the respiratory organs. Attention should be drawn to the presence of polynuclear hydrocarbons in soot from mineral oil combustion because of the interest some of these compounds have from the point of view of carcinogenesis in experimental animals. Although soot is not fly ash, it is thought that such carcinogenic hydrocarbons may well be adsorbed upon or absorbed in fly ash particles.

Apart from this no information concerning the direct influence of fly ash upon the health of man and animals is available. This must be considered a serious lacuna in the present state of knowledge.

As regards the influence of suspended particulates in general, there will probably be no harmful effects observed among sensitive people if the mean level of suspended particulates over the long term remains below about $70\mu g/m^3$, which roughly corresponds to $35\mu g/m^3$ of OECD standard smoke.

Oxides of nitrogen will increase in significance as an indicator of atmospheric pollution. Although they could, in this respect, be ranked with sulphur dioxide, the two can not be compared as regards the predictability of harmful effects on living organisms. Here nitrogen dioxide is of greatest importance. It irritates the mucous membranes, and in higher concentrations may cause irreversible damage to the lungs with immediately fatal consequences (lung edema). Results of recent research indicate that long term exposure to concentrations of NO_2 below 2 mg/m^3 (1 ppm) denatures the lung tissue of test animals (young rats). This process of deterioration can be interpreted as an acceleration of aging.

Biological and Medical Criteria for Air Quality Standards

The issuance of air quality criteria is a vital step in a programme designed to assist the authorities concerned in taking responsible technological, social, and political action to protect the public from the adverse effects of air pollution. The designation of tolerable concentrations of pollutants is thus the result of a multidisciplinary choice in a complex of criteria which are qualitatively and quantitatively dissimilar. But quite apart from the importance one would like to attach to various types of criteria, one should realize that pollution is an evil that cannot be excused and which should be abated and prevented for its own sake.

Because a clean air policy depends in practice, on many more factors than just logic, common sense and good will, only relative, rather than absolute, criteria can be applied. For this very reason, however, one can never be too exhaustive in assessing whether measures to be taken do carry things far enough.

In the matter of multiple choices and decision making, unidisciplinary criteria may serve as guidelines. Such guides are of fundamental importance, and yet all over the world people are engaged in study and evaluation of data that unfortunately are not yet appropriate to the subject in many instances.

There is as yet no agreement as to the interpretability of a number of qualitatively widely differing criteria, e.g. mortality and morbidity statistics versus Pavlovian methods and interpretations. Much depends on philosophy (pragmatism versus dogmatism), and guesswork in a field where knowledge is still lacking. Therefore only a very brief outline is given here, regarding the documents that describe the material in further detail.

There is general agreement that a single, 24-hour exposure of man, animals or plants to a concentration of SO_2 in the order of $200\mu g/m^3$ ($\sim$ 0.08 ppm) will not induce any harmful effect. In a comparable manner long term exposure to a mean SO_2 concentration of $75\mu g/m^3$ ($\sim$ 0.03 ppm) will not result in any damage to man, animals or vegetation. Nevertheless it should be noted that even this relatively low level of SO_2 causes a so-called "desert" of epiphytic lichens and mosses. Higher concentrations and longer exposures will result in a situation where harmful effects become detectable. As regards man and animals such a "sensitivity threshold" may be around 500 to $600\mu g/m^3$ ($\sim$ 0.2 ppm) for a short term exposure (24 hours) and around $150\mu g/m^3$ ($\sim$ 0.05 ppm) as the mean SO_2 concentration over a long period (years). The transition from doses that are harmless to doses that increasingly lead to unwanted and deleterious effects is gradual and depends on the organism and the symptom under study, as well as on environmental conditions. But at exposures to $2000\mu g/m^3$ ($\sim$ 0.8 ppm) for 24 hours, and 300 $\mu g/m^3$ ($\sim$ 0.1 ppm) in the long term one has arrived at doses that induce deleterious effects which are unacceptable from the common sense point of view. These figures apply to SO_2 as an index of pollution by combustion products from the use of traditional fossil fuels.

For a full appreciation of the gradation of this transition from harmless to deleterious exposures, one should realize that a harmful influence may already be acting on single sensitive individuals at concentrations of pollutants which are considerably lower than the level at which exposures begin to have a statistically significant effect on large population groups.

If one really intends to promote clean air in order to prevent unpleasant effects from occurring, one should consider the above mentioned figures concerning a hypothetical "sensitivity threshold" as purely tentative. They hardly give a true picture, and as such they should be appreciated simply as upper limits which in practice should never be reached and which should be rigorously underbid by a policy that intends to quarantee a harmless and comparatively healthy environment based on either the best available knowledge or on qualified guesses.

Bibliography

1. Air Pollution (2nd Edition), Volume I, Air Pollution and Its Effects; Edited by Arthur C. Stern, (Academic Press, New York/London, 1968) par.

11: Effect on the Physical Properties of the Atmosphere; par. 12: Effects of Air Pollutants on Vegetation; par. 13: Biologic Effects of Air Pollutants; par. 14: Effects of Air Pollution on Human Health; par. 15: Effects of Air Pollution on Materials and the Economy.

2. idem, Volume III, Sources of Air Pollution and Their Control; idem, idem, 1968 (par. 50: Air Pollution Control Legislation; par. 51: Air Pollution Standards; par. 52: Air Pollution Control Administration.)

3. Air quality criteria for sulphur oxides; NAPCA Publication No. AP-50, US-DHEW, National Air Pollution Control Administration, Washington, D.C., January 1969.

4. Air quality criteria for particulate matter; NAPCA Publication No. AP-49, US-DHEW, National Air Pollution Control Administration, Washington, D.C., January 1969.

5. Barkman, J. J.: The influence of air pollution on bryophytes and lichens. In: Air Pollution, Proceedings of the First European Congress on the Influence of Air Pollution on Plants and Animals, Wageningen 1968 (Pudoc, Wageningen 1969) 197–209.

6. Barrett, T. W., and H. M. Benedict: Sulphur Dioxide. In: Recognition of Air Pollution Injury to Vegetation: A Pictorial Atlas. Informative Report No. 1, TR-7 Agricultural Committee, Air Pollution Control Association, Pittsburgh, Pennsylvania, 1970: C1–C17.

7. Brasser, L. J., Joosting, P. E., and Van Zuilen, D., Sulphur Dioxide—to what level is it acceptable? Report G300, Research Institute for Public Health Engineering TNO, Delft, The Netherlands, 1967.

8. Frank, N. R., Yoder, R. E., Brain, J. D. and Yokoyama; Eiji; SO_2 (^{35}S labeled) absorption by the nose and mouth under conditions of varying concentration and flow. Arch. Environ. Health 18 (1969) 315–322.

9. Gilbert, O. L.: The Effect of SO_2 on lichens and bryophytes around Newcastle upon Tyne. In: Air Pollution, Proceedings of the First European Congress on the Influence of Air Pollution on Plants and Animals, Wageningen 1968 (Pudoc, Wageningen, 1969) 223–235.

10. Negherbon, W. O., Sulphur dioxide, sulphur trioxide, sulphuric acid and fly ash: their nature and their role in air pollution; Edited by H. N. MacFarland, prepared under EEI Research Project RP62 at Hazleton Laboratories, Inc., for Edison Electric Institute, 750 Third Avenue, New York, N.Y., 10017; published June 1966.

11. Zahn, R.: Über den Einfluss verschiedener Umweltfaktoren auf die Pflanzenempfindlichkeit gegenüber Schwefeldioxyd. Zeitschrift f. Pflanzenkrankheiten (Pflanzenpathologie) u. Pflanzenschutz 70 (2) (1963) 81–95.

ENVIRONMENTAL EFFECTS SPECIFIC TO NUCLEAR POWER PRODUCTION

by *J. A. G. Davids, J. A. Goedkoop* and *M. Muysken*

Summary

The production of electric power from the fission of atomic nuclei is described as far as is necessary for an understanding of the environmental effects. Apart from the discharge of waste heat, common to all thermal power stations, these effects are all reducible to the exposure of man to ionizing radiation. There is broad international agreement on the safe limits for such exposure.

It is shown that the radiation exposure from the normal operation of nuclear power stations can be kept well within these limits. The fission products, which might be released due to malfunction of the nuclear reactor, present a potential hazard, but this can be reduced to virtually zero at acceptable cost. The same applies to the further handling of these fission products during nuclear fuel reprocessing and permanent storage.

Introduction

Bakker and Went, in their opening contribution, mentioned two methods of producing useful energy by reactions involving atomic nuclei: fission of heavy atoms and fusion of light atoms. At present only the former process has been sufficiently developed to constitute a reliable source of energy for power stations and the propulsion of ships. The fission of the uranium or plutonium nucleus is induced by a neutron, and in turn produces further neutrons, enabling a chain reaction to occur. The process also produces fission fragments: atoms with unstable nuclei which eventually, via several intermediate stages, transform into stable atoms. Ionizing radiation is emitted in this radioactive decay process.

In order to obtain useful energy from the fission of atoms several steps are necessary. The uranium must be mined and treated before being fabricated into fuel elements. During this part of the process the nuclear fuel emits only a small amount of radiation. In the next stage the fuel elements are placed in the core of the nuclear reactor. The heat generated by fission is carried away by a coolant which flows through the reactor core past the fuel elements. Part of this heat is transformed outside the reactor into useful energy. At the same time within the reactor core a large amount of radioactive material is formed, primarily the fission products which remain embedded in the fuel elements and are removed from the reactor at the end of the fuel cycle. In addition, the neutrons induce radioactivity both in the coolant itself, and in the corrosion products from the piping system which are carried by the coolant stream. No materials at a distance further than a few metres from the core become radioactive during operation of the reactor. The spent fuel elements, after having been removed from the reactor core, are shipped to the fuel reprocessing plant where the remaining fissionable material, uranium and plutonium, is recovered. The highly radioactive fission products are removed to a permanent storage site.

As will be discussed in the next section, the specific environmental effects of nuclear power are all reducible to the exposure of man to ionizing radiation within internationally accepted limits which are briefly outlined. The normal operation of nuclear power stations and of fuel reprocessing plants as well as accidents with nuclear reactors form the subjects of the following sections.

Environmental Aspects

The production of nuclear power does not lead to the emission of chemical pollutants in the atmosphere such as smoke, soot, fly-ash and the gaseous oxides of sulphur, carbon and nitrogen which result from the combustion of fossil fuel. One environmental effect common to both fossil and nuclear fuels is the discharge of waste heat. According to the second law of thermodynamics only part of the heat energy can be converted into electrical energy, consequently the remainder has to be discharged into the environment. In those reactor plants in operation or being installed, the plant efficiency is about 32% and does not differ appreciably from that of older fossil fuel plants. The remaining 68% is completely rejected to the condenser cooling water. Modern fossil fuel plants have an overall plant efficiency approaching 40%. Approximately 10% of their combustion energy is dissipated directly into the atmosphere by means of the discharge of heated combustion products through the stack, and the remaining 50% is transferred as waste heat to the condenser cooling water. Therefore the present type of nuclear plant needs about 1.7 times more condenser cooling capacity than a modern fossil fuel plant of the same power. It is expected that in future reactor types with a higher core temperature the efficiency of the steam cycle will be increased to between 39 and 43%. For the methods used in dissipating condenser waste heat and for the thermal effects this has on the environment the reader may refer to the contributions of Keller[10] and Koolen.[11]

10. K. J. Keller, "Discharge of Waste Heat," *Electrical Energy Needs and Environmental Problems, Now and in the Future*, Future Shape of Technology Publications, No. 7 (The

In this chapter we intend to evaluate the potential environmental hazard connected with the production of the large amounts of radioactive materials which result from nuclear power generation. The radioactive nuclides which are produced by fission and neutron activation in the reactor core are numerous and have widely differing halflives. They represent many chemical elements and after their eventual release into the environment there are various and sometimes quite complicated routes by which they may reach man. However, their one common factor is that they all contribute to the same environmental effect by increasing man's exposure to ionizing radiation to above the natural background level. Effects on animal and plant life are negligible in comparison.

Radiation Safety Standards

The hazards to health of ionizing radiation were known long before the advent of the nuclear energy industry. Soon after the discoveries of X-rays and natural radioactive substances at the end of the last century it became apparent that unlimited external exposure to ionizing radiation led to acute damage in several organs. It later became evident that such exposure could also induce effects which only appeared after many years. The risk of internal exposure as a result of ingestion of radioactive substances was dramatically demonstrated after the First World War by the high death rate among luminous dial painters who had accumulated large body-burdens of radium and thorium in the course of their work. Finally, in 1927 Muller concluded from experiments with the banana fly Drosophila that exposure of the parents to X-rays caused genetic defects in their offspring.

In many countries these findings led to regulations to prevent over-exposure and, in 1928, the international X-ray and Radium Protection Commission was established to provide guidance and co-ordination. The Commission, the name of which was changed in 1950 to International Commission on Radiological Protection (ICRP), recommends safety standards which are based on biological information accumulated and assessed by experts in the different fields.

The rapid growth of the nuclear energy industry has increased the scope of the Commission's task considerably: safety standards for neutron exposure were required, an assessment had to be made of the internal exposure resulting from inhalation and ingestion of the growing number of new radionuclides and the possibility of people being irradiated non-occupation-

Hague: Future Shape of Technology Foundation, April, 1971), pp. 17–24; this book pp. 45–54.
11. J. L. Koolen, "Biological Effects of Cooling Water Discharge," *Electrical Energy Needs... (op. cit.,* note 10), pp. 39–49; this book pp. 63–76.

ally, due to release of radioactive nuclides into the environment, had to be considered.

The most recent basic recommendations of ICRP were issued in 1966.[12] They give dose limits for two groups of individuals: adults exposed in the course of their work and members of the public. The dose limits of the second group, which are one-tenth of those of the first group, are given in Table 8.

The ICRP has based these limits on the assumption that for late-appearing effects, notably the induction of malignant tumours, no threshold dose exists below which the risk is zero. An upper limit of risk has been estimated by a linear extrapolation of the dose effect relationship for high doses administered at high dose rates, down to the dose limits mentioned in Table 8. The ICRP estimated that the lifetime risk of cancer from a single whole-body exposure of 1 rem has an upper limit between 10^{-5} and 10^{-4}. The true risk figure is somewhere between zero and the upper limit.[13] A more accurate direct determination is, however, beyond the scope of the available scientific methods.

If a large proportion of the population were exposed the consideration of hereditary effects would play a major role in setting the dose limit to the gonads. The recommendations of the geneticists as to dose limitation would be met in practice if, averaged over the population, the dose in the gonads during the first 30 years of life were limited to 5 rems. Again it is assumed that the hereditary effects are linearly related to the gonad dose, that no threshold dose exists, and that the dose effect relationship is independent of dose rate.

Table 8. Dose limits for members of the public (ICRP 1966)

Organ or tissue	Dose limits [a]
gonads, red bone marrow (whole body uniformly irradiated)	0.5 rem in a year
skin, bone	3 rems in a year
thyroid (for age above 16 years)	3 rems in a year
thyroid (for age below 16 years)	1.5 rems in a year
hands and fore-arms, feet and ankles	7.5 rems in a year
other single organs	1.5 rems in a year

a) The rem is a measure that includes an estimate of the biological effectiveness of different types of ionizing radiation. For X-, γ- and β-rays 1 rem corresponds to an energy absorption of 100 ergs per gram of tissue. 1 millirem (mrem) = 10^{-3} rem.

12. ICRP, *Recommendations of the International Commission on Radiological Protection*, Publication No. 9, (London: Pergamon Press, 1966).

13. ICRP, *The Evaluation of Risks from Radiation*, Publication No. 8 (London: Pergamon Press, 1966).

In view of the uncertainty of the existence of a threshold dose, the ICRP recommends keeping exposure to ionizing radiation as low as practicable. Therefore the risks involved in its use have to be balanced either against the benefits obtained or against the risks inherent in alternative methods. The ICRP dose limits are not intended to include doses from medical, diagnostic or therapeutic, exposures or doses from natural background irradiation. The latter is due to natural radioactive substances in the body (about 20 mrems/year), to cosmic rays (about 30 mrems/year at sea level) and to natural radioactive substances in the soil. This last component varies considerably depending on the soil type. It ranges from about 20 mrems/year over sedimentary rocks to about 170 mrems/year in granite districts. Above highly radioactive soils, in the Kerala State in India for instance, dose rates of up to 4.000 mrems per year have been measured.

On the other hand, the ICRP dose limits apply not only to the production of nuclear power but also to a number of other human activities which may entail an increased exposure to ionizing radiation. For instance, flying at an altitude of 10 km gives an increased dose rate due to cosmic rays of about 0,5 mrem per hour. Watching colour television closely may expose one to soft X-rays. Radioactive substances are used for various non-medical purposes. Even the choice of construction materials for buildings can mean a difference in indoor dose rate of about 20 mrems per year.

In planning the release of radionuclides into the environment the expected dose people receive from both external and internal irradiation has to be assessed. Internal exposure may result from ingestion of contaminated water and food or inhalation of contaminated air. Therefore the ICRP has issued figures derived from the dose limits[14] laying down the maximum permissible intake, by either route, of many radionuclides. The maximum permissible intake of a single radionuclide is a complex function of its physical characteristics, such as type of emitted radiation and radioactive half-life, and its chemical form, which determines its pathway within the body. In Table 9 some figures have been assembled which serve to illustrate that, depending on the radionuclide, the maximum permissible intake (in microcuries) may differ by about a million.

14. ICRP, *Report of Committee II on Permissible Dose for Internal Radiation*, Publication No. 2 (London: Pergamon Press, 1959); and *Report of Committee IV on Evaluation of Radiation Doses to Body Tissues from Internal Contamination Due to Occupational Exposure*, Publication No. 10 (London: Pergamon Press, 1968).

Table 9. Maximum permissible continuous daily intake by either inhalation or ingestion for single radionuclides. (figures for members of the public)

Radionuclide	Critical organ[a])	Ingestion μ Ci[b])	Inhalation μ Ci
hydrogen-3 (tritium) as oxide	total body	8	3
ruthenium-106	gastro-intestinal tract	0.03	0.06
cesium-137	total body	0.03	0.04
strontium-89	bone	0.02	0.02
iodine-131	thyroid (adult)	0.005	0.006
strontium-90	bone	0.0008	0.0007
radium-226	bone	0.00003	0.00002
plutonium-239	bone	0.008	0.000001

a) Dose in this organ is limiting.
b) The curie (Ci) is the unit of radioactivity equal to 3.7×10^{10} disintegrations per second. μ Ci $= 10^{-6}$ Ci.

Nuclear Power Stations in Normal Operation

As mentioned in the introduction, the bulk of the radioactive materials formed by fission remain embedded in the fuel elements, which are usually bundles of metal tubes containing sintered oxide fuel. During fabrication the fuel elements are inspected very carefully in order to ensure that no leakage of radioactive materials from the fuel element will occur during operation of the reactor. However, in practice it is not always possible to achieve 100% leak-tightness. It must therefore be expected that the coolant will be slightly contaminated by fission products. A second source of contamination of the coolant is the activation, during the passage through the core, of impurities. These impurities are mainly the result of corrosion of the piping system through which the coolant passes.

The first nuclear reactors built in the US and the UK were cooled either by air, which was discharged through a stack, or by water, which was returned to the river. The coolant is now circulated in a closed or almost closed circuit, and is either water, carbon dioxide or helium. For future reactors liquid metal (sodium) and molten salt mixtures (fluorides) are envisaged. In most cases the heat is transferred from the primary coolant to a steam cycle. Exceptions are the boiling water reactor and the gas-cooled reactor with direct cycle gas turbine. With liquid metal cooled reactors it is at present considered necessary to have an intermediate cycle between the coolant flowing through the reactor and the steam cycle.

Most reactors being built in the Netherlands, Belgium and Germany belong to one of the two types of water cooled reactors, either the pressurized water reactor (PWR), shown schematically in Figure 7 or the boiling water reactor (BWR), shown in Figure 8. As these reactors are the most prominent at present, and will continue to be so for at least the next decade, more detailed discussion will be restricted to them.

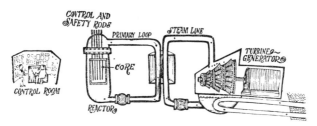

Figure 7. Pressurised water reactor (PWR) power plant (from Atomic power safety, US AEC, DTI 64-62700).

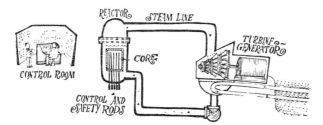

Figure 8. Boiling water reactor (BWR) power plant (from Atomic power safety, US AEC, DTI 64-62700).

In the PWR the coolant system is completely closed. A certain part of the coolant is continuously bypassed through the purification system, consisting of filters and ion exchange beds which trap all the circulating radioactive substances except those which are gaseous. The gases released from the reactor system are led through filters and charcoal beds in order to remove radioactive particles and iodine. The radioactive isotopes of noble gases can not be removed in this way. After a delay, in which most of the radioactivity decays, they are released to the environment through a stack, at concentration levels below the limits set by the authorities and based on the standards discussed in the third section. Used filters and ion exchange beds constitute the main body of solid radioactive waste from the power plant and are removed from time to time to a burial ground or other storage site.

Unavoidably, there is a certain amount of waste water in which minor amounts of radioactive substances are dissolved. This low-level liquid radioactive waste is usually discharged into the cooling water stream from the condenser in order to dilute it to below the limits allowed for discharge to the environment.

In a modern PWR plant the low-level waste from the purification system will contain mainly tritium. This is formed by neutron absorption in the boron which is added to the primary coolant for additional control of the reactor.

With the BWR system the primary coolant of the reactor is transformed to steam inside the reactor vessel, and is then led directly to the turbine. The steam next passes to the condenser and is condensed to feedwater, which is led back to the reactor vessel. A continuous extraction of vapour from the condenser is required in order to ensure sufficient vacuum. This results in higher release of radioactivity from the stack than is the case with PWR reactors. As an illustration, Table 10 gives annual release figures for a modern power station of each type:

These figures show that it is possible to operate both types of plants with radioactivity releases which are only a fraction of the official limits.

Table 10. Actual and permitted releases of radioactivity in curies per year.

Location Type	San Onofre, Cal, USA 450 MW PWR		Gundremmingen, W.-Germany 250 MW BWR	
	actual	permitted	actual	permitted
through stack: noble and activation				
gases	260	567,000	17,500	1,920,000
iodine-131	0.0001[a])	0.8[a])	0.17	22
liquid effluent: fission and corrosion				
products	8	[b])	2.4	14.4
tritium	3500	[c])	24	4800

a) includes radioactive particles
b) not given, actual average concentration was 14% of permitted value
c) not given, actual average concentration was 0.2% of permitted value

Fuel Reprocessing Plants and Waste Storage

Spent fuel elements are reprocessed in order to separate the valuable uranium and plutonium from the fission products. All processes used today are of the aqueous type, which consists essentially of first dissolving the fuel in a nitric acid solution and then extracting uranium and plutonium by means of an organic liquid. The aqueous solution of fission products, which is highly radioactive, is further concentrated by evaporation and stored in shielded stainless steel tanks which are cooled continuously and monitored for leaks. The whole process yields, apart from the high-active concentrated solution, a relatively large volume of liquid waste with a much lower radioactivity content in addition to gaseous radioactive waste, notably krypton-85 and tritium.

The permanent storage of the high-active waste has been a matter of concern for many years. The most promising solution seems to be to solidify the waste products and store them in caves in geological structures remote from ground water, preferably in salt deposits, which give, according to geologists, an absolute guarantee that the waste will not contaminate the biosphere. Two methods for solidification have been developed: melting the fission products together with SiO_2 into a glassy substance, or calcination at high temperature resulting in a ceramic powder which is then packed in metal cylinders. In the USA new reprocessing plants will be equipped with a solidification unit, while, in Kansas, a permanent storage facility is being installed in a salt deposit.

A high purification of the medium and low-level liquid waste resulting from reprocessing is technically feasible. The extent to which it is put into practice depends, among other things, on the local disposal capacity. For instance, disposal of liquid radioactive waste into coastal seawater is used on a substantial scale in the English reprocessing plant at Windscale on the Irish Sea. The disposal capacity was first carefully assessed by dye dilution experiments and by considering all possible routes via which human exposure could result. The limiting factor appeared to be the concentration of the fission product ruthenium-106 by a red sea-weed Porphyra, which is harvested in the area and transported to Wales for consumption by a small number of people. It is estimated that the Windscale discharge might give, in a group of about 100 people, an exposure to the intestine of up to 50% of the ICRP dose limits.[15] Disposal of fission products in coastal water is also carried out in the French plant at Cap la Hague near Cherbourg. On the other hand the Eurochemic plant at Mol, and several plants in the USA,

15. H. J. Dunster, A. W. Kenny, W. T. L. Neal, and A. Preston, "The British Approach to Environmental Monitoring," *Nuclear Safety*, Vol. 10, No. 6 (1969), 504.

purify their liquid waste to a higher degree. The latter method produces of course more solid radioactive waste, which has to be disposed of in a selected burial ground.

Krypton-85 has the longest half-life (10.8 years) of the noble gas nuclides produced by fission. It easily passes the filter systems, which retain particles and iodine, and is discharged entirely into the atmosphere. The resulting exposure in the vicinity of reprocessing plants is at present negligible, but with the anticipated growth of nuclear power the estimated world-wide distribution by the year 2000 would give an appreciable dose, mainly on the skin, as a result of external exposure. It is therefore of interest that in the Oak Ridge National Laboratory a rather simple purification method has recently been developed in which krypton and xenon are absorbed in cooled freon. The krypton removal efficiency is claimed to be better than 99.9%. This method will not only solve the future krypton-85 problem but will also be tested for its applicability in BWR plants in order to reduce the emission of short-lived radioactive noble gases.[16]

The production of tritium (half-life 12.3 years) in the cooling medium and moderator of various reactor types has already been mentioned. Tritium is also formed in the fuel as a ternary fission product at a daily rate of about 50 mCi per MW electric power. With stainless steel, but not with zircaloy, some diffusion through the cladding of the element occurs. The tritium in the fuel is released in the reprocessing plant. With the reprocessing methods currently in use, about 75-99% of tritium present in the fuel appears as tritiated water (HTO) in the medium and low-level waste. It is released into the environment either as tritiated water vapour in the atmosphere, or directly into surface water. Because of the low radiotoxicity (see Table 10) of tritium as HTO the local capacity for disposal in surface water is considerable.[17]

16. W. G. Belter, *Recent Developments in the U.S. Low-level Radioactive Waste Management Program-An Overview for the 1970's*, Presented at the IAEA-ENEA Symposium on Low and Intermediate-level Waste, Aix-en-Provence, September 7-11, 1970.

17. H. T. Peterson, J. E. Martin, C. L. Weaver and E. D. Harward, "Environmental Tritium Contamination from Increasing Utilization of Nuclear-Energy Sources," *Proceedings of the Environmental Contamination by Radioactive Materials Seminar*, Vienna, March 24-28, 1969 (Vienna: International Atomic Energy Agency, 1969), p. 35; and D. G. Jacobs, "Sources of Tritium and Its Behaviour upon Release to the Environment," Presented to the U.S. Congress, Joint Committee on Atomic Energy, *Hearings, Environmental Effects of Producing Electric Power*, Part 1, Appendix 5, (Washington: U.S. Government Printing Office, 1969).

Accidents in Nuclear Power Stations and Their Consequences

Before a nuclear power station is built an extensive report on the safety of the station is drawn up and submitted to the authorities for approval. This safety report considers in detail the amounts of radioactivity which may be released to the environment during normal operation, as well as the amounts which would be released should various possible or hypothetical modes of plant malfunction occur, or should serious damage be done to parts of the plant.

The fission products in the fuel constitute the main body of radioactivity in an operating reactor. A far smaller amount is present in the coolant of the primary system. Malfunctioning of the plant or damage to the primary coolant system such that only an enlarged leakage of primary coolant results, will not have serious consequences for the environment. The amounts of radioactivity involved are relatively small and the purification system and the filter system in the stack for trapping the gaseous radioactivity have sufficient capacity to cope with larger quantities than arise during normal operation. However, the release of fission products from the fuel must be prevented as completely as possible. There are several barriers incorporated in the reactor plant which prevent their release. In the first place a large part of the fission products is non-volatile, and thus remains in the fuel. The gaseous fission products will escape from the fuel, but are retained inside the cladding. Thus, as long as the fuel rods remain intact, no fission products will escape to the surroundings.

The next barrier is the primary coolant system. Then again the reactor and its primary coolant system are contained in a leak-tight building. Relatively small leakages from the primary cooling system and from the containment building must be accepted, but can be handled in such a way that release to the environment is less than the imposed limits.

A more serious situation can only occur if the cladding of the fuel rods fails disastrously. A conceivable cause for such a failure is overheating followed by melting of the cladding. This again could be a result of insufficient cooling of the fuel rods or of a sudden power surge. The most serious case is a large rupture in the primary cooling system, leading to a loss of the coolant from the system. As this occurrence cannot be completely ruled out in the safety evaluation of the plant, the containment building is designed to cope with such a sudden rupture, followed by partial melting of the fuel cladding. An independent safety measure is the provision of an emergency cooling system, which comes into operation automatically to prevent melting of the cladding. Should the worst happen, and the emergency cooling system not operate fast enough to prevent melting of the cladding, a larger release of radioactivity to the surroundings than accepted for normal operation must be taken into account. However, such an extreme case has never yet occurred, and is only a hypothetical assumption in order to obtain design figures for the containment building and emergency cooling system. The other possible cause for melting of the fuel rods, a sudden power surge, is even more unlikely than a sudden large rupture in the primary coolant system, at least with the present types of power reactors. However, one such accident did happen in 1961 with the SL1, a prototype army package power reactor in the USA. This incident was presumably caused by jerking a control rod out of the core by hand, and cost the lives of three people. The amounts of radioactivity released to the environment, in this case the Idaho desert, were approximately:

iodine-131   80   curie
cesium-137   0.5 curie
strontium-90   0.1 curie

It must be added that the SL1 reactor was built in a corrugated iron shed; there was no containment building.[18]

In the reactors of power stations now in use the neutron absorbing capacity of the individual control rods is kept so small that there would be no melt-down of the fuel if one rod were suddenly withdrawn from the core. Furthermore, the control systems shut the reactor down if the power or the rate of power increase exceed preset limits. And last, there is a self-limiting mechanism, also operating in the case of the SL1 incident, which reduces the power level after a rise in temperature of the fuel, or the coolant, or both. The discussion of this inherent safety feature would lead outside the context of this paper.

More radioactivity was released in an accident at Windscale with an early reactor which was cooled by air. The air was released from the stack after passing through the core. Thus there was no closed primary coolant system nor a containment building. In October, 1957 part of the fuel elements in this reactor melted. It has been estimated that the following amounts of radio-activity were released from the stack:

iodine-131   20,000 curie
tellurium-132   12,000 curie
cesium-137   600 curie
strontium-89   80 curie
strontium-90   2 curie

The spread of radioactivity could be measured with highly sensitive instruments as far away as Frankfurt, Germany. However, the only practical measure which was necessary to protect the population was to prohibit the

18. J. R. Horan and W. P. Gammill, "The Health Physics Aspects of the SL-1 Accident," *Health Physics*, Vol. 9 (1963), 177.

use of milk in an area of about 200 square miles for 25 days in the greater part of this area, and for 44 days in the most highly contaminated region. Fourteen members of the staff received a dose of over 3 rems in the period of 13 weeks, the highest dose being 4.7 rems. Some of the population in the downwind sector received a thyroid dose as a consequence of the iodine, the highest recorded being 16 rems in a child. An independent committee set up by the Medical Research Council concluded "that it was in the highest degree unlikely that any harm had been done to the health of anybody whether a worker in the Windscale plant or a member of the general public."[19]

Conclusion

The first manifestations of nuclear energy in the form of atomic weapons were so awe-inspiring that its civil applications have been surrounded with great care for public health. As a result, it can be said with confidence that the normal operation of nuclear power plants presents no environmental problem apart from the heat release which is common to all thermal power stations. Care must be taken, however, to contain the large amounts of radioactive fission products, within the reactor, even in the case of a malfunction, and afterwards during reprocessing and long-term storage. Technically this is feasible, and at acceptable costs.

19. J. F. Loutit, W. G. Marley, and R. S. Russell, "The Nuclear Reactor Accident at Windscale"-October 1957: "Environmental Aspects," Appendix H, *The Hazards to Man of Nuclear and Allied Radiations*, a *Second Report to the Medical Research Council* (London: Her Majesty's Stationery Office, 1960).

Chapter V

ENERGY NEEDS AND ENVIRONMENTAL PROBLEMS:
UNITED STATES POLICY*

by *Clarence Davies, 3d*

THE OVERALL SETTING

Fragmented Government

To understand United States energy policy one must first realize that currently there is no such policy. There is, rather, a wide variety of partial, fragmentary, often contradictory policies related to energy.

The absence of a coherent energy policy results to a great extent from the fragmented, pluralistic nature of the American Government. Governmental authority is divided between the Federal Government and the 50 State governments; the Federal Government is divided among the Congress, the Judiciary, and the Executive Branch; and the Executive Branch is divided among a large number of bureaus and agencies which sometimes ignore the policy leadership of the President and his political appointees.

The Constitution of the United States gives certain powers to the Federal Government and leaves all remaining authority to the States. Over the past forty years, through liberal judicial interpretation of the Constitution, the Federal powers have significantly expanded and the Constitution is no longer a major obstacle to new Federal activities. However, the States remain a vital part of the political system. The Congress is elected on a State basis and, most importantly, the two major political parties have their power base in the State and local governments. In fact, there are no national parties —there are two major coalitions of local parties. The States carry out a wide variety of major functions, including many related to energy policies.

The Federal Government consists of the Congress (House and Senate), the Federal Judiciary, and the Executive Branch. Action by both the Congress and the Executive Branch is necessary for the successful promulgation and implementation of any policy, but there are few mechanisms to bring the two institutions together. The Congress in recent years often has been controlled

* The views presented herein are those of the author and do not necessarily represent the views of the U.S. Government or any agency thereof.

by the opposite party from the President's, and even when it is controlled by the same party, loyalty to the party label has been so weak that the President has had considerable difficulty in getting his proposed programs approved.

Given the diffusion of power and the limited ability of the President to control the fate of his recommendations, including his budgetary recommendations, the heads of the operating bureaus and agencies are to some degree forced to fend for themselves. Their primary reliance has been on outside interest groups which can exert pressure across the divided institutions at the Federal level. Thus the operating agencies accumulate significant power bases of their own, and are highly responsive to the outside interests which give them political viability.

The pattern of power within the bureaucracy makes it most difficult to achieve a unified policy on matters which affect several agencies. The Department of the Interior often reflects the views of the oil and coal industries, the Atomic Energy Commission is responsive to the needs of the nuclear industry, and the Federal Power Commission must consider the requirements of the electric utilities. Those at the higher levels who would reconcile these diverse interests often do not have the political muscle to force an agreement. A unified energy policy runs contrary to the diffuse nature of the political system.

The Energy Industry

Just as it is erroneous to talk about the existence of an energy policy, it is also misleading to talk about an energy industry. Although there exist Federal anti-trust statutes which forbid collusion within an industry, the leaders of the oil companies, for example, share a common interest, are subject to the same regulatory authorities, and do talk to each other. There is little such sharing between the oil companies and the coal companies, or between either of these and the utilities. The notable exception is when the same company has interests in several different fuel sources, a situation which is becoming increasingly more common.

The dominant private sector forces in the production of energy are the fuel companies. The oil companies in particular represent one of the strongest political forces in the nation. Petroleum accounts for only 13% of the fuel used to generate electricity (Table 1), but it is the major overall source of energy in the U.S., and the oil companies also tend to play an important role in the production of natural gas.

Almost half of the fuel consumed by electric utilities is coal. However, politically the coal companies have had little influence, except in the few states which are economically dependent on coal mining.

A comparison between the coal and oil industries with respect to their relations with government illustrates at least one significant point about the operation of the American system. The oil industry is one of the most regulated industries, whereas there are comparatively few regulatory authorities over coal. As indicated above, the oil companies are far more influential than the coal companies. Given the rhetoric of free enterprise capitalism, and its antagonism to government controls, this might seem like an anomaly. But in fact the rhetoric is often belied by practice, and the major corporations have rarely hesitated to use the government and its authority to foster their goals. The regulations on oil promote market stability, keep out "undesirable" competition, and serve a number of other purposes beneficial to the private sector. The line between public and private in the U.S. is not a sharp one.

Since energy policies are largely built around fuels policies, it is necessary to explore the particular policies which relate to each of the fuel sources used in the generation of electricity. Section II describes the nature of the industries involved and the government policies with respect to these industries.

Table 1. Gross Consumption of Energy Resources by U.S. Electric Utilities, 1970, by Major Sources.

Source	Trillion BTU	Source % of total Utility Consumption	Utility Consumption % of total source use
Anthracite Coal	48	—	25
Bituminous Coal & Lignite	7,776	46	57
Natural Gas, dry (excl. LNG)	4,025	24	18
Petroleum	2,263	13	8
Hydropower	2,674	16	100
Nuclear	208	1	100
Total	16,967	100	25 (av.)

Source: Div. of Fossil Fuels, Bureau of Mines, U.S. Department of the Interior.

Coal

The history of coal in the U.S. over the past 50 years has generally been a history of loss of markets to other competing fuels. In 1920 bituminous coal production reached 569 million tons a year. It declined until World War II, reached a peak of 630 million tons in 1947, and then declined again. However, beginning in 1960 coal production began a rather remarkable recovery. Preliminary 1970 figures indicate that production will reach 590 million tons.[1]

Coal is still the largest single source of fuel for electric utilities, and the recent recovery of the coal industry has largely been due to the great increase in electricity generation. However, two other factors have changed the nature of the coal industry and contributed to its revitalization. One was the decision in the early 1950's of the United Mine Workers Union to cease its attempts at protecting employment in the industry and to cooperate with the mine owners in making production more efficient. The result was a decline in the number of miners employed from 442,000 in 1949 to 134,000 in 1965 with little loss in production.[2] The Mine Workers' decision also brought some degree of stability to the unstable and often violent labor-management relations which had characterized the coal fields.

The second major change in the industry was the purchase of the larger coal firms by large diversified corporations, particularly oil companies. The ten largest coal producers mine 65% of the coal, and of these ten only one is an independent coal company. Two of the three largest coal producers are owned by oil companies, and five of the largest ten are. In the past five years, oil companies have increased their share of coal production from 7% to 28% of total coal production.[3]

Coal is the only major energy source not subject to direct public regulation with the exception of health and safety regulations. Production is not publicly controlled, and exports and imports are not subject to any special restrictions. The price of coal at the minemouth fluctuates with the market. However, the price to utilities is greatly influenced by the cost of transportation, and rail rates for coal are regulated by the Interstate Commerce Commission (ICC). In recent years the ICC has been quite responsive to

1. U.S. Dept. of the Interior, Bureau of Mines, "Production of Mineral Fuels and Hydropower in the U.S. Since 1800," pp. 20–22.
2. Resources for the Future, *U.S. Energy Policies – An Agenda for Research* (Washington, 1968). p. 70.
3. Beck, Laurence D., and Stuart L. Rawlings, *Coal: The Captive Giant* (Washington, 1971), p. 11.

the needs of both the railroads and the coal producers. This has resulted in highly favorable rates for coal transportation, notably for the "unit train," a series of specially designed cars which shuttle back and forth between a mine and a power plant.

The government assists coal production in other ways. The producers enjoy tax benefits in the form of a depletion allowance which allows 10% of the gross value of coal produced to be deducted before computing income for income tax purposes (up to a limit of 50% of net income). The same advantage is given to producers of other mineral resources, although the rates vary depending upon the type of mineral. The depletion allowance for coal is significantly less than that allowed for oil, so that the allowance actually results in a competitive disadvantage to coal producers. The Federal Government also funds some research on new uses of coal and improved methods of mining and processing. The dollar value of the research is small compared with what is expended for nuclear power research (see Table 2), but is quite large compared to industry research expenditures.

Environmental and safety regulations have had a profound impact on the coal industry. The Federal Coal Mine Health and Safety Act of 1969 has already resulted in increased costs to the industry, and its impact will probably increase in future years. The environmental effects of strip mining have been subject to increasing criticism. The States have begun to pass more stringent laws regulating the way in which strip mining can be conducted and requiring restoration of mined areas. In 1970 the Nixon Administration submitted to Congress proposed Federal legislation which would authorize the Federal Government to directly regulate strip and underground mining in those States which did not have adequate regulations or which did not enforce the regulations on the books. The water pollution effects of coal mining, particularly the problem known as "acid mine drainage," also have been subject to increasing Federal and State regulation.

The environmental regulations which have had the greatest impact on coal have been those dealing with air pollution. U.S. air pollution control officials regard sulfur dioxide as the air pollutant most responsible for adverse effects on human health, and the burning of coal by electric utilities accounts for about 40% of the sulfur dioxide air pollution.[4] Over the past few years, progressively tighter standards have been set for the percentage of sulfur allowed in coal—New Jersey and New York City now limit the sulfur content to 1% and will shortly be limiting it to less than 0.3%; a number of localities have placed limits of 2–3% on sulfur content; and the Federal

4. Estimate derived from U.S. Dept. of Health, Education, and Welfare, Public Health Service, Environmental Health Service, "Nationwide Inventory of Air Pollutant Emissions – 1968," pp. 9–11.

Government recently proposed a national air quality standard of 80 micro-grams of sulfur dioxide per cubic meter (arithmetical mean) which for many communities will mean a sulfur content limitation of 1 % or less. The available supply of low-sulfur coal is probably not adequate to meet the demands produced by existing and proposed air pollution regulations. Private industry and the government have therefore been investing large sums in developing the technology to remove sulfur oxides from the stack gases. A number of processes are now being tried, and there is little doubt that some will prove technically and commercially feasible. But the timing of such development is critical, and unless the technology to control sulfur oxides emissions from both new and existing power plants is available within the next few years there will be a severe decline in the demand for coal. Coal already has lost a major part of the important East Coast market to residual oil because of the demand for low-sulfur fuels.

Oil

The major American oil companies are a powerful political force. The oil industry is comparatively concentrated—in 1965, the twenty largest companies accounted for 85% of domestic refining capacity and produced 60% of all domestic crude oil.[5] The largest companies are vertically integrated, controlling the supplies of crude oil, the pipelines to the refineries, the refineries, and the retail outlets for the finished products.

The supply of crude oil is subject to a unique and complicated set of government regulations. The key actors in the regulation of domestic supply are the States. Over the years, the major oil-producing States have established boards or commissions which set monthly limits on the total amount of oil which can be produced within the State, formulas for allocating the total amount among the individual producers, and rules for reservoir development governing the number and spacing of wells. The basic factor responsible for the development of these State regulations is Anglo-American property law which gives ownership of subsurface minerals to the owner of the surface property. The reservoirs of oil overlap the surface property lines, and thus without State regulation there was no way of preventing a property owner from drilling a well and draining the oil from his neighbor's property.

The Federal Government, under the Connally ("Hot Oil") Act of 1935, is committed to enforcing the State regulations by preventing the movement in interstate commerce of oil produced in violation of State law. However such enforcement has not been necessary, because the industry has welcomed the market stability provided by the State controls.

5. Resources for the Future, *op. cit.*, p. 25.

Imports of oil have been controlled since 1959 by Federal regulations. Imports account for slightly more than 20% of the petroleum products consumed within the U.S.[6] The ostensible reason for the imposition of import quotas has been "national security," but the quota system has been a major factor in controlling the price of oil. Not only do the quotas limit the total supply of fuel available and thus support the price structure, but they also shift costs among the major refiners. The import quotas are allotted to domestic refiners, but the inland refiners process only domestic crude and thus do not use their quota. Instead, they trade them to East Coast refiners in exchange for domestic oil and a considerable profit. The quota system has been characterized as "a profit-sharing scheme which forces East Coast refiners to share the benefits of low-cost imports with inland refiners."[7] The import system has been under considerable pressure because of the demand, particularly on the populous East Coast, for low-sulfur oil to meet air pollution control regulations. Thus, at the present time, the East Coast is allowed to import almost unlimited amounts of residual oil, resulting in a marked decline in the use of coal.

The Federal Government also controls oil supply through its ownership of oil-bearing lands, including the offshore Outer Continental Shelf. The government has generally followed a policy of encouraging development of oil production on these lands, through leasing of mineral rights to private corporations, and about 10% of all domestic oil now comes from public lands.[8] The one area where development has not been encouraged up until now is the shale oil deposits in Colorado, Utah, and Wyoming. In these three states, the shale rock is estimated to contain the equivalent of close to 600 billion barrels of recoverable oil, or enough to supply the domestic consumption needs of the U.S. at current rates for more than 100 years. The Federal Government owns more than 70% of the shale oil land containing close to 80% of the oil.[9] In 1930 it withdrew all Federally owned oil shale land from leasing. However, in his Energy Message of June 4, 1971, President Nixon directed the Secretary of the Interior to expedite the development of an oil shale leasing program and to proceed with such a program if environmental concerns can be satisfied. The privately owned oil shale has not yet been commercially developed, but it is not certain whether this is more because of the economics of development itself or because of the effects on existing crude oil prices and facilities, since the major oil companies are also the major owners of the privately held shale lands.

6. U.S. Dept. of the Interior, Office of Oil and Gas.
7. Resources for the Future, *op. cit.*, p. 44.
8. *ibid.*, p. 47.
9. *ibid.*, p. 116.

Oil receives a far more favorable tax treatment than coal. The Federal depletion allowance for oil is 22%, the highest for any mineral. Also, intangible drilling expenses, which represent the major fraction of the capital cost of wells, are permitted to be charged as current expenses instead of being written off over the life of the wells. This decreases current income taxes during the period of development and gives command over a volume of cash funds that may be thought of as interest-free deferred taxes.[10]

Environmental concern over oil production and use has focused primarily on water pollution from oil spills and on air pollution from oil combustion. Amendments to the Federal Water Pollution Control Act passed in 1970 provide severe penalties for spills occurring during the drilling, transportation, or storage of oil. The Act provides that, except for a few minor exceptions, the owners of vessels which spill or discharge oil in violation of the Act shall "be liable to the United States Government for the actual costs incurred... for the removal of such oil by the United States Government in an amount not to exceed $100 per gross ton of such vessel or $14,000,000, whichever is lesser" except when the discharge was willful, in which case the owner is liable for the full amount of removal.[11]

The imposition of air pollution regulations controlling sulfur oxides emissions has resulted in significantly increased investments for the oil industry. The major firms have invested several hundred million dollars in facilities to remove sulfur from oil. These added costs have been reflected in higher oil prices. However, the industry has gained by capturing a portion of the market previously held by high-sulfur coal.

Natural Gas

Natural gas is now the second largest fuel source for electric utilities, being exceeded only by coal. Natural gas is generally found in the same locations as oil, its production is closely associated with oil production, and consequently the major oil companies are also the primary producers of natural gas. However, unlike oil, the natural gas industry is not vertically integrated. The transportation of the gas from the producing fields is undertaken by independent pipeline companies. It is then sold "at the city gate" to utility companies.

The movement of natural gas is subject to public regulation at every stage, from well to consumer. The States apply production limitations of the same type as are applied to oil. Under the Natural Gas Act of 1938, the construction of new pipelines and extensions and abandonments of existing

10. *ibid.*, p. 40.
11. Federal Water Pollution Control Act, sec. 11 (f) (1).

pipelines must be approved by the Federal Power Commission (FPC), and the Commission regulates the price which the pipelines can charge to the distributing utilities insofar as the gas is shipped interstate.

Until 1954, the FPC took the price charged the pipelines by the gas producers as a given part of the pipeline's costs which could not be regulated by the FPC. However, in that year the Supreme Court ruled[12] that under the 1938 statute the FPC was also required to regulate the price of gas at the wellhead, if the gas was shipped interstate. The manner in which these prices should be regulated and whether the FPC should regulate them at all have been matters of intense controversy ever since. The Congress has twice voted to free field prices from regulation, but the bills were vetoed by two Presidents.[13] The FPC has not been able to arrive at a regulatory system completely satisfactory to itself, and in March 1971 it published an order exempting small producers from price regulation.[14]

The price which the local gas utilities charge to customers is regulated by State utility commissions, in much the same way that the price of electricity is regulated.

The degree to which the Federal Government controls the price of gas is greater than for any other major fuel. The FPC has generally followed a policy of limiting the distribution of natural gas and of keeping the price comparatively low. It has limited distribution because of a concern for the limited supply. The price has been kept low in part as an outgrowth of the restrictions on distribution and in part because of the Commission's desire to protect the ultimate consumer. Thus while the government regulation of oil generally serves to keep the price of oil high, the FPC's regulation of gas keeps the price of natural gas low. The FPC's regulations have also discouraged interstate shipment of gas because of the markedly higher price which can be obtained for gas produced and consumed within one state.

The adverse environmental effects of natural gas are minimal, and the industry has been aided by the recent flurry of pollution control regulations. The restrictions on the sulfur content of coal and oil have resulted in a number of utilities converting to the use of natural gas.

Nuclear

The development of nuclear energy for the generation of electricity has evolved out of the Federal Government's efforts during World War II to develop nuclear weapons. Unlike any other portion of the energy industry,

---

12. Phillips Petroleum Co. v. Wisconsin et. al., 347 U.S. 672.
13. Resources for the Future, *op. cit.*, p. 61.
14. FPC Order No. 428, March 18, 1971.

nuclear power began as a government monopoly.

In the years since 1946, when the civilian Atomic Energy Commission (AEC) was created, the government has turned over more and more aspects of nuclear power development to private industry. The ultimate policy goal is to have all aspects of nuclear power generation in private hands, subject to governmental regulation. Currently, the only major function still performed entirely in government facilities is fuel enrichment. The AEC operates three large gaseous diffusion plants for enriching uranium, and the Nixon Administration has proposed selling these facilities to private operators. However, the proposed sale has become highly controversial. It is not known who, if anyone, would buy the plants, whether there would be competition for their ownership, and what the true worth of the plants is. The AEC and the Joint Congressional Committee on Atomic Energy have both tended to oppose the sale of the plants.

The divestment by the government of ownership of the elements of nuclear power production does not mean that the government's influence is not felt in the development and pricing of nuclear power. In fact, the field is so permeated by government involvement that it is difficult to tell whether nuclear power production is really economically competitive, given the variety of Federal subsidies and regulations.[15] Imports of uranium have, in effect, been embargoed for the past 7 years. The government has accumulated large stockpiles of uranium, and this has influenced the domestic price. Facilities for uranium enrichment are still AEC-owned. The AEC licenses each major stage in the construction of nuclear power plants, and attaches detailed conditions to its licenses. It provides partial insurance against accident for nuclear plants. Finally, it exercises close control over fuel reprocessing plants and radioactive waste disposal, and itself operates facilities for the disposal of high-level wastes.

The primary forces in the government's nuclear "establishment" are the AEC and the Joint Congressional Committee on Atomic Energy. The relationship between these two groups is unique because to a great extent it bridges the division between the Congress and the Executive Branch. The Joint Committee, the only major substantive committee which includes members of both the House and Senate, has taken a strong hand in the internal policies and administration of the AEC, and thus in most cases the views of AEC and the Committee are identical. This is not the case with most Executive Branch agencies and the Congressional committees responsible for their activities.

As of December 31, 1970 there were 16 nuclear power plants in commercial

15. See Resources for the Future, *op. cit.*, pp. 101–109.

110

operation, with a generating capacity of 6,493 Mw.[16] Estimates of the future use of nuclear energy vary widely, and the orders for new plants have fluctuated considerably (see Figure 1). The AEC estimates that by 1980, 25% of electricity generation will be from nuclear plants.

Figure 1. % Nuclear of Total Generating Capacity Ordered, by Year.

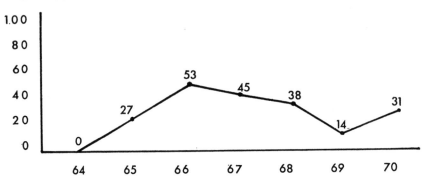

Source: AEC, "Central Station Nuclear Plants," Jan. 1, 1971, p. 3.

The major environmental problems from nuclear power production are: 1) emission of low-level radioactive wastes; 2) disposal of high-level radioactive wastes; 3) plant safety; and 4) thermal pollution, i.e., the discharge of heated cooling water into streams.

The emission of radioactive wastes into air and water is no longer a serious problem with respect to nuclear reactors, although the public often views it as a problem. The plants built to date emit some tritium and krypton into the atmosphere, but the amounts are so small as to be unimportant in comparison with other variables determining human exposure. The amounts of radioactive material going into the water are also very small. In 1970, the private manufacturers of reactors announced the availability of waste disposal systems which would significantly reduce escape of radioactive materials from a nuclear plant, and these systems are likely to become standard equipment on new plants.

The disposal of radioactive wastes is perhaps the most difficult environmental problem associated with nuclear power. The problem is particularly acute for long-lived high-level wastes. The AEC has promulgated a policy

16. U.S. Dept. of the Interior, Office of Power Statistics.

requiring the conversion of such wastes from liquid into solid form and has proposed storing them in an abandoned salt mine in Kansas. Although elaborate precautions would be taken to insure the safety of the storage facility, the people of Kansas have not taken kindly to the idea, and it is uncertain whether the project will be approved.

Plant safety has been of considerable concern to the American public. Although there has been some controversy over safety among those knowledgable about nuclear energy, it has not been considered a major problem. This may change, however, as the breeder reactor comes closer to being a reality and as the size of conventional reactors continues to increase.

Nuclear power plants discharge approximately 50% more heated water than conventional fossil fuel plants. There has been increasing controversy and concern over the possible effects of thermal discharges, and utilities are now being forced to make investments for the construction of cooling towers. However, much research remains to be done on the diffusion and effects of heated water from power plants, and the degree to which the siting and costs of nuclear plants will be affected by thermal pollution considerations is not yet clear.

Hydroelectric Power

The proportion of electricity generation accounted for by hydroelectric power has been declining steadily. To a great extent this is because the most economical sites for dam construction have been already used. There has also been increasing opposition on the part of environmentalists to the building of dams and consequent flooding of large areas of land. Generation of electricity through pumped storage hydropower has been used increasingly by utilities as a method for meeting peak-hour demands, although it still accounts for less than 10% of hydropower generating capacity and is actually a net user of electric power.

Three Federal agencies construct multi-purpose water resource projects which include power generation facilities—the Army Corps of Engineers, the Department of the Interior's Bureau of Reclamation, and the Tennessee Valley Authority. The government retains ownership of these facilities, and as of 1968 there were 137 Federally owned hydroelectric power plants with a developed capacity of 20 Mkw. These plants accounted for 43% of total developed conventional hydroelectric capacity.[17]

Hydroelectric projects developed by private industry or by the States and localities must be licensed by the Federal Power Commission. The FPC

17. Federal Power Commission, "Hydroelectric Power Evaluation," (March 1968), p. 1.

licenses generally expire at the end of 50 years, and the first of these are now coming up for renewal. This has raised in a new form what has been the major question in the whole hydroelectric power area, namely what should be the role of the Federal Government. The private utilities supported by some members of Congress have urged the Federal Government to get out of the power generation business. The Executive Branch position has varied depending upon the views of the President in power, but in recent years there has been a general policy of maintaining the mixed public-private *status quo*.

Research on New Technologies

As is true with all aspects of U.S. fuels policy, there is no coordinated effort to fund Federal research on fuels technology. Questions as to what projects should be funded and what the level of expenditures should be usually are decided within each category of fuel according to the political power of the public and private institutions involved, taking into account the structure of the industry. Little attempt is made to view fuels research as a whole or to decide priorities among the different fuel sources.

Table 2. Federal Funding for Fuels Research (in $M).

Fuel	FY 1970	FY 1971 (est.)	FY 1972 (est.)
Coal	44	68	93
Oil	6	6	6
Gas	—	—	—
Nuclear	214	219	232
Hydro	<1	<1	<1
Other	1	1	1
Total	266	295	333

Source: Derived from the Budget of the U.S. Gov't., FY 1972 Appendix. The figures represent rough estimates.

The dominant portion of Federal fuels research funds are devoted to nuclear energy. The emphasis is on the development of the Liquid Metal Fast Breeder Reactor, and more than $ 100 million a year is being spent to develop the breeder reactor. Research on fusion reactors, the next stage of development after the breeder, is at a lower level, averaging about $ 30 million a year. Some environmentalists have urged that higher priority be given to development of a fusion reactor because they believe it would pose fewer environmental problems than the breeder, but a careful comparative analysis of the environmental effects of the two types of reactors has yet to be done, and much of the difference in expenditures between the two types is due to the more advanced stage of development of the breeder.

About 25% of Federal coal research is directed at the development of economically feasible methods for coal gassification or liquification. The industry has also invested in this research, and several pilot plants have been built for converting coal to a liquid or gas. The figures in Table 2 include under coal approximately $ 25 million in 1972 for research on removing sulfur oxides. Stack gas removal is essential if coal is to maintain its market in those areas with stringent air pollution controls. The increase in coal research has been due to increased expenditures for sulfur removal techniques and for coal mine health and safety.

The President's June 4 Energy Message contained three priority research commitments: completion of a successful demonstration of the liquid metal fast breeder reactor by 1980; a doubling of Federal support for sulfur oxide control demonstration projects in Fiscal Year 1972; and an expansion of the coal gassification program. The President also promised increased support for a variety of other energy research projects.

The small amount of funds devoted to research on oil primarily reflects the ability of the oil industry to finance its own research efforts and the absence of any obvious technological problems faced by the industry.

There is little research on new methods of producing energy, such as batteries, magnetohydrodynamics or solar energy, because of the absence of any political force to support such efforts and the lack of any government agency with responsibility for undertaking such research. Research is the function which suffers most from the absence of any centralized or coordinated energy policy and which most obviously shows the fragmented nature of the government's efforts in the energy field.

GENERATION AND TRANSMISSION POLICY

Electricity in the U.S. is generated and transmitted by several different types of organizations. (see Table 3) Three-fourths of the power is generated by private investor-owned companies. However, the largest number of systems are non-Federal public systems owned and operated by municipal governments. The Federal Government itself accounts for 13% of the power generated, and the Tennessee Valley Authority, a Federal body, is the largest single electric power system in the country. Finally there are non-profit rural electric cooperatives, most of which are financed by low-interest Federal loans.

The Federal Power Commission regulates the wholesale rates of electricity moving in interstate commerce. The courts have given a broad interpretation of this authority, and, given the increasing interconnection of all electric systems, the Commission's authority could become the prime determinant of electric rates. However, this has not yet happened, and at the present time the State public utility commissions are the key regulatory authority. The State commissions regulate the rates and services of the private systems and, in some States, of the municipal systems. As with most State functions, the performance of the utility commissions has been mixed. Their job has been made comparatively simple because the costs of power generation, at least until recently, have steadily declined.

Table 3. Components of U.S. Electric Power Industry.

Ownership	Number of Systems (1968)	% of Total Power Generated (1969)
Private	405	76
Public-Non Federal	2,075	10
Cooperative	960	1
Federal Government	40	13
Total	3,480	100

Source: S. D. Freeman and J. F. Weinhold, "Policy Alternatives for Resolving the Power Plant Siting Problem" in *Environmental Aspects of Nuclear Power Stations*, International Atomic Energy Agency, Vienna, 1971, p. 759.

The environmental effects of electricity generation and transmission cannot be separated from the effects of the various fuel sources which have been discussed above. One exception is the concern over the appearance of and amount of land consumed by transmission lines. At some point costs of underground transmission may become low enough so that placing wires above ground will no longer be necessary. But this point is at least 10–20 years away, unless added funds are provided for research on the technology of underground transmission.

The construction of a number of new power plants has been halted or delayed by public concern over the air and water pollution effects of the plant. The siting of power plants has become a major public issue in many areas. In an attempt to bring about a greater degree of reconciliation between environmental concerns and the need for increased power generating capacity the Nixon Administration in February submitted to Congress legislation requiring advanced planning for power plant siting. The bill would require long-range planning by all electric utilities, with continuous 10-year projections of power needs and additional required facilities for generation and transmission. Annual reports containing such projections and information related to activities of each utility to protect environmental values would be made readily available to the public. Advance review of tentative sites for power plants would be held 5 years prior to commencement of construction with authority by a State or regional reviewing agency to approve alternative sites as environmentally more suitable. Finally, a preconstruction review would require each utility to apply for certification of a bulk power plant or major transmission line 2 years prior to commencement of construction.

Consideration of environmental factors during the earliest planning stages for power plants and transmission lines will be necessary, whether or not the power plant siting bill becomes law. It will be necessary if environmental goals are to be achieved, and it will be necessary if adequate power is to be provided to meet the nation's needs.

The incorporation of environmental factors into the construction and operation of power generating facilities will increase the cost of power to the consumers. This is consistent with national environmental policy which is that the cost of goods and services should include *all* costs of their production—including the cost of preventing damage to the environment.

ENVIRONMENTAL POLICY

The environment has become a major concern for the majority of Americans. This has occurred only within the past few years, and the reasons for the upsurge of public concern are no easier to pinpoint precisely than the reasons for other mass shifts in public opinion. One reason is certainly that pollution and other environmental problems were getting worse. The deterioration in environmental quality was accompanied by increasing affluence and leisure time, which made people more aware and more affected by such problems as water pollution. Another factor which cannot be disregarded is the desire for a unifying political issue to relieve the divisive pressures of Vietnam and the urban Blacks.

The public concern has resulted in the passage of many new laws at the local, State, and Federal level; significantly increased spending for pollution control; and new organizations, both public and private, to deal with environmental issues. At the Federal level, a Council on Environmental Quality was created in the Executive Office of the President to advise the President on environmental policy and to coordinate the actions of Federal agencies with respect to actions affecting the environment. The existing pollution control programs and legal authorities were consolidated in a new Environmental Protection Agency (EPA). The EPA has almost 6,000 employees, and its proposed budget for Fiscal Year 1972 is almost $2.5 billion.

Public policy has relied primarily on the establishment and enforcement of environmental standards, particularly air and water pollution standards. Under the 1970 amendments to the Clean Air Act, the Federal Government is authorized to set national ambient air quality standards. The States will set emission standards from particular sources, but the State standards are subject to Federal approval, and the Federal Government itself will set emission standards for major new industrial plants, including power plants. Up to now, air pollution standards (with the exception of motor vehicle emission standards, which are Federally established and enforced) have been set jointly by the States and the Federal Government, the States proposing air quality standards and the Federal Government approving them.

Water quality standards have been established through the same State-Federal pattern that previously prevailed in air pollution. Amendments to the Water Pollution Control Act have been proposed which in many respects parallel the changes made last year in the Clean Air Act but which would retain the Federal-State process for setting water quality standards. The coverage of the standards and the authority for direct Federal enforcement would be significantly increased.

The Federal policy has been to place primary reliance on the States for enforcement of the standards. The effectiveness of the State in carrying out their enforcement responsibilities has varied widely from State to State, but there are serious deficiencies in the great majority of State programs. The Federal Government does have some authority to enforce standards directly, but up until now this authority has been used primarily to prod the States into taking action. The 1970 amendments to the Clean Air Act, the government's resurrection of an 1899 water pollution statute, and the proposed amendments to the Federal Water Pollution Control Act all provide the Federal Government with much stronger enforcement authority, and it is likely that in the future much more of the burden of enforcement will be borne by Washington.

The government has also recently shown interest in using economic incentives as a supplement to pollution control standards and enforcement. Last year the Administration proposed a tax on lead in gasoline, designed to encourage the availability and marketability of low- or non-leaded gasoline. The lead tax has been re-submitted to the Congress this year and the President has proposed an even more significant measure for charges on emissions of sulfur oxides. The sulfur oxides emission charge would be levied on sulfur emitted into the atmosphere from combustion or distillation of fossil fuels and from other possible sources. To the extent that sulfur is removed from fuels, or from the stack gas, no payment of the charge would be required.

The use of economic incentives is an attempt to correct existing market forces which encourage firms to pollute. Air and water are usually free goods, and controlling pollution simply adds to the cost of the products produced. Also, given the major role of the States in setting and enforcing environmental standards and the resulting variations in stringency of controls, a firm which installs pollution control equipment may be put at a significant competitive disadvange with respect to rival firms located in more lenient States. The use of Federal taxes to put a real cost on pollution provides a means for making it economically desirable for the private sector to control pollution. However, such taxes are viewed as a supplement to regulatory actions, not a substitute for them.

The lead tax and sulfur charge represent an attempt to make private sector polluters pay for the costs of damage caused by their pollution. The Executive Branch of the Federal Government has established such payment as a general policy goal. The Congress is more inclined to have the government itself pay part of the costs of pollution damage through tax benefits to firms installing pollution control equipment. This policy was reflected in the 1969 amendments to the tax code, which provided that firms were entitled to a rapid tax write-off on the cost of pollution-control equipment installed. Many States also provide tax relief for industries installing pollution control equipment.

The electric utilities and the suppliers of fuel have been more affected by environmental policies than any other segment of the economy. To the extent that there is a conflict between economic growth and environmental quality, the generation of electricity is the cutting edge of such conflict.

Air pollution control regulations governing sulfur oxides have had a major impact on the market for coal and oil and have increased the price of fuels for the electric utilities. Such regulations have also forced the utilities to make major investments in control devices, such as scrubbers and precipitators to remove particulates. The regulations have encouraged a shift to nuclear power generation, but this has been balanced by increasing concern over the environmental effects of nuclear power plants. Water pollution regulations dealing with thermal discharges may have a major impact on the cost of generating power and on the siting of power plants, particularly nuclear plants.

As discussed at the beginning of this paper, there is no integrated U.S. energy policy. There are elements of an environmental policy, but even in the environmental area it has only been within the past few months that the threads of a cohesive policy have begun to knit together the air pollution, water pollution, radiation, and other environmental programs.

There is no mechanism for bringing together the disparate elements of energy policy with the various aspects of environmental policy. The energy policy makers generally depend upon a fuel-oriented constituency for their political existence. Thus the view of the energy policy makers tends to be much like the view of the energy makers—environmental concerns are an obstacle to be overcome. Pollution regulations and pressures from conservationists are viewed as somewhat unpredictable factors which interfere with the normal price-setting mechanisms, which distort normal market demands, and which obstruct the planning of capital investments.

The view of the environmental policy makers is often equally narrow. The implications of pollution control regulations for energy supply and demand are often not considered, and, when they are considered, it is usually with inadequate data and sometimes with an attitude that economic considerations are not the proper concern of those responsible for saving the environment.

Recent developments give some hope that a more coherent and comprehensive view of the interrelationship between energy and environment can be achieved. The Office of Science and Technology in the Executive Office of the President has had a small staff which has provided policy advice on overall energy matters and which is keenly aware of environmental considerations. The Council on Environmental Quality is beginning to turn its attention to the environmental aspects of energy policies.

Perhaps the most important development has been the initiation of the so-called "102 process." Section 102 of the National Environmental Policy Act of 1969, the act which created the Council on Environmental Quality, states that any agency of the Federal Government planning to undertake an action which will have "a significant environmental impact" must file a statement with the Council describing the action and its impact and also describing the alternatives which were considered by the agency. These statements are circulated for comment to relevant Federal, State, and local agencies, and the final statements and comments are considered public documents. The effect of this process has been to force the energy decision-makers into considering the environmental implications of their actions before such actions are taken. It has also injected the environmental agencies, such as the Council and the Environmental Protection Agency, into the decision-making process on such matters as licensing nuclear power plants or constructing hydroelectric dams.

There is a growing awareness both within and outside the government that neither energy policy nor environmental policy can be successful if they proceed down totally independent paths. Production of fuels and electricity to meet the ever-growing demands of the U.S. economy will be ruinous to the environment if environmental impacts are not adequately considered.

The imposition of environmental regulations without due regard for the energy needs of the economy will result in severely distorting and handicapping the efficient development and use of energy resources and will, in the end, prove self-defeating.

A number of attempts have been made over the past few years to bring some coherence to Federal energy policies. The pressure for a government body which could consider all aspects of energy policy has increased as environmental concerns have begun to erode the stability of the fuels industries and the electric utilities.

In 1967, President Johnson directed his Office of Science and Technology (OST) to make a thorough study of energy resources. The Congress denied the Executive Branch adequate resources to conduct such a study, but OST has maintained a small energy staff which has done a remarkable job of trying to bring some unity of thought to the government's disparate energy policies.

In August 1970, President Nixon named a committee of his cabinet-level Domestic Council to coordinate government energy policies. The committee includes representatives of 12 Federal agencies. It has made a number of recommendations to the President, many of which were included in a special Presidential message on energy sent to Congress in June.

The President also has endorsed the creation of a new Department of Natural Resources, and he submitted legislation in 1971 to form such a department. The new agency would include an energy component which would have central responsibility for energy research. However, the proposed department would leave most of the scattered Federal regulatory authority untouched—the Federal Power Commission would continue to regulate gas, hydroelectric power, and electricity rates, and the Atomic Energy Commission would retain much of its influence over nuclear power. There is also some question as to whether Congress will approve the creation of the department, although the Congress itself has been considering the creation of a Joint Committee on Energy.

All of the governmental efforts to unify energy policy must fight an uphill battle against the centrifugal tendencies of the political system. The private sector may achieve unification before the government does. The oil companies and a few other large firms have begun to turn themselves into "energy companies" with large interests in oil, gas, coal, and uranium. Whether the government will make a concerted effort under the anti-trust laws to prevent such consolidation is not yet clear, but even if it does it will be fighting powerful forces in the private sector.

Technological developments have encouraged private sector consolidation and will probably continue to do so. The gassification or liquification of coal will make coal, the most abundant U.S. fuel, interchangeable for all purposes with oil and gas. The development of the breeder and fusion nuclear reactors will require large capital resources for plant construction. It will also accelerate the current trend, resulting from improved high-voltage transmission technology, toward ever-larger power plants. One can envision the energy field dominated by a handful of major energy companies, supplying all fuels, and a few giant utilities, controlling major generating facilities and the transmission networks.

Whether the increasing consolidation of the private sector will provoke a corresponding trend within the government is uncertain. It is not known, for example, whether the companies which now have interests in several fuel sources will support the governmental efforts to coordinate or bring together the Federal energy agencies. It seems likely that their fear of losing influence over the regulatory process will lead them to oppose any change in the status quo. However, the advantages of dealing with a fewer number of government agencies and the hope of additional Federal research funds may weigh more heavily than the fear of more stringent regulation.

Environmental policy will certainly undergo major changes in the coming years, but the field is so volatile that predictions are dangerous. The decision of the Congress rejecting funding for the construction of a supersonic transport is perhaps a good indicator of the future of environmental politics. The decision represented a major shift in public thinking, because for the first time an issue of technological development became a matter of widespread controversy and the key question was posed of even if we *can* accomplish a major technological breakthrough do we *want* to accomplish it. The fact that the Congress answered in the negative, despite strong Administration pressures, illustrates the depth of concern over the relationship between technology and the good society.

On the other hand, the issue was also the first time that industry and the labor unions combined forces against the environmentalists, and this pattern will probably be repeated as the economic impact of environmental concerns becomes increasingly apparent. So far as we now know, there is no scientific or technological reason for believing that increased power production and improved environmental quality are incompatible. The basic questions are economic and political. Will the society be willing to pay the added costs involved in meeting environmental standards? Will the government have the political will to enforce such standards? If the answers to these questions are positive we can have adequate supplies of electricity and a good environment.

Chapter VI

THE BASES OF THE U.K. POLICY FOR ELECTRICAL ENERGY PRODUCTION*

by *Craig Sinclair*

Introduction

Publically enunciated and articulated policies for the development, production and distribution of electrical energy in the United Kingdom are of comparatively recent origin and are still very much in course of definition. The power necessary to develop Britain's manufacturing capabilities in the last century and the early part of the present relied overwhelmingly on indigenous supplies of coal. With the development of central electrical power stations they became prime users of coal and it is only within the last fifteen years that other fuels, oil and uranium, have made substantial inroads into the supply of coal for electrical power generation.

The U.K.'s growing dependence on imported fuel arises partly from the physical inability of the coal suppliers to match the economy's rising electrical energy requirements, but mainly because imported oil has become relatively cheaper. Thus, while imports of oil contribute towards the problem of the balance of payments, they represent effectively lower energy costs for manufacturing, and exporting, industries. Latterly the concern over pollution and in the wider context, amenity, has become a more important factor in policy making with regard to electrical energy. Thus in short summary, the three main strands of national policies for electrical energy are derived from,

a) the existence of a large but declining coal industry, giving rise to social and regional problems,
b) a growing supply from overseas of economically attractive fuel oil, presenting problems of security of supply and foreign exchange, and
c) a certain degree of pressure on amenities and natural resources of a growing electrical supply network.

* The author wishes to thank the individuals within the industry and government departments who have assisted him in preparation, and Mr. John Surrey of the University of Sussex' Science Policy Research Unit who made a large contribution to the final form of this chapter.

In the immediate post-war years after 1945, the main and perhaps only problem was to find fuel, essentially coal, to satisfy the rapidly increasing demands for energy of a general industrial recovery. The concern to develop coal production sufficiently to meet this demand was an entirely domestic problem. By the early fifties, it was realised that coal production could never be expanded sufficiently to meet demand, and consequently a greater reliance was placed on oil and the nuclear power programme had its inception. Even under the Conservative administration in the 1950's and early 1960's government policies with regard to fuel were by no means non-interventionist. Mr. Richard Wood, in 1962, defined the objectives of Conservative fuel policy: "I should not like to dictate to consumers what fuel would be best for them. Guaranteeing a certain target for the coal industry would have a very bad effect, and a complete freedom of fuel use would also have very serious consequences. Nor am I at all sure that the solution of allowing everybody to use the cheapest fuel they want is the right one. If we become too dependent on foreign supplies we endanger our own security—we have to take the middle course." After 1964, against a background of cheaper fuel abroad and a protected U.K. coal industry whose rationalisation had gone some way but had not solved the problems of competition in price, and with a considerable, established, nuclear power programme, Command Papers were published in 1965[1] and 1967[2] dealing directly with Fuel and Energy Policies. These policy statements were delivered amidst a degree of independent comment and political discussion.[3]

Demand Forecasts

It is first pertinent, however, to examine briefly demand forecasts for electrical energy. There has never been any lack of forecasts of energy demand and supply, which is remarkable in the light of errors revealed in earlier forecasts by the subsequent course of events. The Industrial Inquiry made in preparation for the National Plan was at the time the most comprehensive attempt yet made. The Plan, in assuming a national growth of 25% between 1964 and 1970, implied that this 3.8% annual increase would (utilising a complex relationship between economic growth and energy consumption)[4]

1. *Fuel Policy*, Command Paper No. 2798 (London: Her Majesty's Stationery Office, 1965).
2. *Fuel Policy*, Command Paper No. 3438 (London: Her Majesty's Stationery Office, 1967).
3. P.E.P., *A United Kingdom Fuel Policy* (London, 1967); Lord Robens, *The Future of Coal*, A paper presented to the British Electrical Power Convention, 1965; G. Tugendhat, "Freedom for Fuel," *Institute of Economic Affairs* (London, 1963).
4. *The Financial and Economic Obligations of the Nationalised Industries*, Command Paper No. 1337 (London: Her Majesty's Stationery Office, 1961).

result in a increased coal equivalent demand of $3.8 \times 11 \times 10^6$ tons. The resulting total inland energy demand was 337 million tons of coal equivalent, analysed in *Table 1*. The estimates were assessed by the then Ministry of Power from the returns made by the fuel industries and the electricity and gas industries. These assessments were the subject of discussion by the Energy Advisory Council (see below) and the final result is given in *Table 2*. The major point of interest emerging from the discussions was the Government's (1965) acceptance that it would be imprudent to rely on a coal market in 1970 greater than 170 to 180 million tons including exports. The Chairman of the National Coal Board, Lord Robens, a former Labour Cabinet Minister, had publicly declared that he accepted the economic challenge for coal and that coal could face competition without protection; however, he considered 200 million tons as the minimum for a viable coal industry, a position which he appeared to cling to when the forecasts were again substantially revised downwards as a result of the review of fuel policy in 1967.

Long-term forecasts are, of course, of greater uncertainty, but are necessary in the context of setting background for consideration of policy. In *Table 3* forecasts up to 1985[5] are given by extrapolation of both the Plan's predicted rate and from the regular forecasts carried out by the independent National Institute of Economic and Social Research. A review of prospects given in the 1967 White Paper revised the estimates as summarised in *Table 4*. The reduction was principally due to considerations relating to the slower growth of the British economy than postulated in the 1965 National Plan, recognition of the need to incorporate substantial quantities of natural gas from the North Sea into the future fuel economy, and the continuing problem of coal.

Organisational Structure

Of the four major fuel industries in the U.K. three are nationalised bodies, coal, gas and electricity, while oil is in the private sector. The Ministry of Fuel and Power Act, 1945, imposes the responsibility on the Minister, of "securing the effective and co-ordinated development of coal, petroleum and other... sources of fuel and power in Great Britain, and of promoting economy and efficiency in the supply, distribution and consumption of fuel and power, whether produced in Great Britain or not." This responsibility can be discharged using three main policy tools:
a) general powers, e.g., fiscal, import controls, etc. as used by Governments to exercise regulation on the economy as a whole,

5. P.E.P., *op. cit.*

b) powers, specific to fuel production, e.g., the regulation of the production of oil and natural gas in U.K. parts of the Continental Shelf, and

c) the Minister's own powers in relation to the nationalised industries, for example in appointing Boards and some wide, if generally limited in usefulness, opportunities of control under the Nationalisation Acts. In respect of pollution and amenity considerations, for example, the electrical industry is required to take into account the impact of policies of siting and transmission of electricity on the local environment."

The Minister, since 1970, in the Department of Trade and Industry in this respect, has responsibility together with the Ministry of the Environment for several aspects of amenity relating to electrical power production—the siting of power stations, nuclear and fossil fuelled natural gas shore terminals, coal tips and waste disposal in the sea, problems of noise from gas turbines, transformers and 400 kV switching and, of course, oil in the sea. Not all of these relate solely to electrical energy production and some have strong links with other aspects of Government policy. For example, the establishment and siting of oil refineries in the U.K. has implications for the flow of foreign capital into the country, the production of a variety of petrochemical feedstocks and permits a diversity of sources of oil to be used.

Consultation and discussion at various levels ensures Ministerial influence. The financial investment and borrowing programmes of the three industries are reviewed annually against medium-term national economic and energy forecasts so that the development and capital spending for the five years ahead is laid before the Ministers each year. The Ministers, of Power (or since 1970 the Secretary of State at the Department of Trade and Industry) for England and Wales, the Secretary of State for Scotland—in consultation with the Treasury and economic departments—fix upper limits on investment during succeeding years.

In order that an adequate return on both past and future investment is secured, the Boards set quinquennial financial and economic objectives. These were originally put forward under the 1961 White Paper,[6] Financial and Economic Objectives of the Nationalised Industries, and amended under the 1967 White Paper, Nationalised Industries: A Review of Economic and Financial Objectives.[7] Programmes of capital investment are also scrutinized for broad consistency with each other. Under a "gentlemen's agreement," first introduced in wartime, the National Coal Board gives the Minister prior notice of price increases; similar agreements exist with gas and electricity. Some criticism of this arrangement has been made by the Commons

6. The Financial and Economic..., op. cit.

7. Nationalised Industries: A Review of Economic and Financial Objectives, Command Paper No. 3437 (London: Her Majesty's Stationery Office, 1967).

Select Committee on the Nationalised Industries and it was partly in response to this criticism and the feeling that the nationalised industries had been free from real commercial discipline that the 1961 White Paper laid out the financial objectives.

Co-ordination in another field, that of research and development, is ensured in planning, execution and application by the Ministers' Advisory Council on Research and Development (ACORD) operating with an independent chairman and with members drawn from all fuel industries, from industry generally, from the universities, Government Departments and the Atomic Energy Authority. This group considers the adequacy of research programmes and the general balance between different lines of research, co-operation and overlap. The terms of reference are given more fully as:

a) To advise the Secretary of State for Trade and Industry on research and development in relation to his statutory duty of securing the effective and co-ordinated development of coal, petroleum and other sources of fuel and power in Great Britain, and of promoting economy and efficiency in the supply, distribution, use and consumption of fuel and power, whether produced in Great Britain or not;

b) To advise the Secretary of State of new scientific and technical knowledge or applications of knowledge throughout the world, which in the opinion of the Council should be taken into account in the performance of his statutory duties;

c) To keep the whole field of fuel and power under continuous review with the object of identifying problems needing research and development and advising the Secretary of State of these problems with a view to discussion with the industries.

Further, the Research Liaison Committee joins the N.C.B., C.E.G.B. and the Gas Council's representatives.

On the wider issues of national fuel policy the Minister acts as chairman of the Co-ordinating Committee which consists of the chairman of the National Coal Board, Electricity Council and the Gas Council. Finally, in order to bring into consultation on fuel policy, not only the heads of nationalised industry, but also the leaders of the oil industry, industry as consumers, and the trade unions, representatives of these interests together with Government officials meet "to consider and advise the Minister about the energy situation and outlook and the plans and policies of the fuel and power industries in relation to national objectives for economic growth." The nationalised industries published detailed annual statistical accounts but in general little detailed policy. Neither are the full ranges of "profitability of various sectors of the industry in question fully discussed, for example in relation to tariff policy. The Select Committee, formed of members of all parties, is thus, though only inadequately supported in terms of technical secretariat, a useful watchdog on performance.

Under the various Nationalisation Acts, Consumers Councils are designed to ensure the industry is in some measure accountable to the customer. There is a widespread feeling that they are impotent—except perhaps at local level— particularly when attempting to assess the difficult problem of value for money in the fuel market, since insufficient economic and technical data is available to them.

Structure of the Electricity Industry

In England and Wales a "federal" group of boards has operated since 1958. A single national generating board transmits electricity for sale to a number of regional distribution boards who in turn supply customers. These three classes are represented on the co-ordinating Electricity Council. Additionally, around one-ninth of generation in England and Wales is by private industrial users for their own use. In Scotland two vertically integrated boards generate, transmit, and sell electricity directly to the consumer. The Electricity Council is reponsible for the central co-ordination, in England and Wales, of policy, particularly that relating to finance, research and industrial relations and it is comprised of a chairman, two deputies and three other independent members, three C.E.G.B. representatives and the regional area board chairman. In recent years, partly due to the overwhelming power in the industry of the C.E.G.B. and its ability to pass on its costs to the area boards via the Bulk Supply Tariff, this structure has received considerable criticism, and the previous government was legislating to revise this structure when it left office.

Main Objectives of Fuel Policy in Respect of Electricity Production

Policies specific to the electric energy production process must take account of the more general considerations of the policies to be pursued in respect of all forms of energy, primary coal, gas and nuclear power. The Civil Service structure is best described with reference to *Diagram 1.*

The separate fuels or industries are controlled by separate sections under Deputy Secretaries who to some extent are open to the criticism of acting as a mouthpiece in Whitehall for the industries they "sponsor." In the 1965 White Paper on fuel policy, the overriding consideration was that the energy sector should contribute maximally to the economy and to the balance of payments. The full range of objectives in British fuel policy are listed below, and it will be seen that one of the main problems in formulating policies has been the incompatibility, under certain circumstances, of the policy objectives:

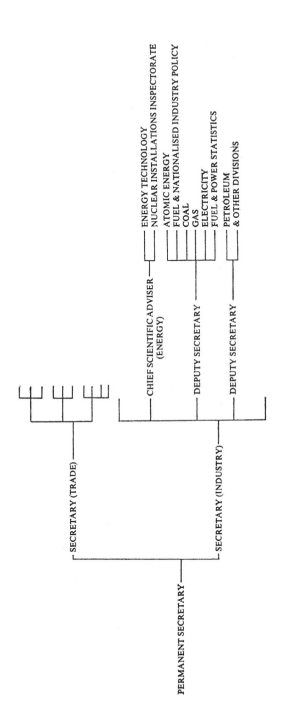

a) Adequate and continuous supplies of fuels of suitable quality should be available to sustain the desired rate of growth in the economy.

b) The price of fuels should be such as to enable them to play their part in making the United Kingdom economy as a whole competitive, particularly in relation to other countries of Western Europe.

c) The fuel industries should be technically progressive and viable.

d) Imports of fuel, particularly of oil, which on any reckoning will grow fast, should be in the form that is least costly in terms of foreign exchange.

e) Consumer freedom of choice, apart from being desirable in itself, is an essential guide to the efficient planning of supplies, provided that the prices paid by consumers fully reflect all the relevant costs.

As mentioned in the introduction, the protection of the indigenous coal industry has been a major concern since the post-1945 period and to this has been added concern over security of foreign oil supplies, competition from oil and nuclear power. Latterly natural gas has become a primary fuel competitor and increased public concern with social costs and amenity is being expressed by private pressure groups such as the Clean Air Society and Council for the Conservation of Rural England.

Coal

The social costs of the protection of the coal industry have loomed large in all discussions of fuel policy over the last decade. With regard to electricity production, policies for coal are clearly important as electricity is the major user of this fuel, taking an increasing percentage of the industry's output. In 1964 coal accounted 82% of total usage, oil 12% and nuclear energy 4% with hydro electricity making up the remainder. The C.E.G.B. in 1960 took 26% of the coal output, 36% in 1964, the proportion growing to almost 50% by 1970. In 1964, 180 of the 233 stations were coal fired, and bought coal from 450 mines. In 1965, fifteen coal fired stations were under construction; however, by 1971 no further coal fired units were being planned. On the 1965 estimates nuclear power was seen as being competitive by the early 1970's a little over 10% cheaper than coal and a little under 10% cheaper than oil. While these estimates have had to be revised, adversely from the position of nuclear power, the effect seems simply to be a delay in the end of coal fired station building by a year or two.

In recent years the optimum rate of contraction of the coal industry has been hotly debated, for there is obviously a limit above which the benefits of a faster rundown are offset by rising social costs. The protection which is afforded the coal industry has been reviewed from time to time as outlined below, but is governed by the general considerations that protection:

a) should not be such as to make British fuel costs uncompetitive as com-

pared with those of our main competitors in Western Europe;
b) should not be so great as to prevent the development of an adequate and well-balanced oil-refining industry in the United Kingdom;
c) should be sufficient to prevent the market for coal from declining faster than output can, in practice, be economically reduced, and to help the industry to bring about smoothly its transition to a more compact and competitive structure.

Methods of protecting the coal industry—described in 1964 as the largest employer in the world after General Motors with around 500,000 employees and supporting 28 Members of Parliament—as laid out in the 1965 and 1967 Papers are:
a) the duty on heavy oil (and other oil used for burning) at a level of about $6 per ton and a levy—a revenue duty— on road transport and fuels;
b) a ban on the import of coal (American or Polish);
c) a ban on the import of Russian oil;
d) ad hoc, supposedly short-term, measures, including administrative measures to increase coal burned in power stations. (This extra coal use by the CEGB and the gas industry due to these administrative measures to give further protection to coal cost the taxpayer £20m. ($48m.) over the two financial years, 1968 and 1969.)

Successive Governments have set themselves against a continuing general direct subsidy for coal production as being inconsistent with the idea of self support and efficiency and have called for a coal industry which at least breaks even, remembering that massive sums of fixed capital in the coal industry have been written off in recent years. No Government, however, has found itself able to reduce or abolish the coal-protecting $6 a ton oil tax originally introduced as a revenue tax, or permit free imports of coal.

Social Problems

The need for concentration of the coal industry has resulted in very substantial manpower reduction over the past ten years, which is presently around 30,000 per annum. The redundancies are exacerbating the situation in already high unemployment regions—Scotland, the North East and South Wales, and policies have had to take account of this fact. An expansion in Government Training Centres was made and employers were given financial assistance in training and retraining programmes. Special funds are being made available to assist redeployment and resettlement in other industries and to facilitate early retirement of coal miners. These amounted to some £45m. by 1970–71. The regional development policies of the Labour Government were used to alleviate unemployment in these regions and consisted of grants ranging from 25% for the establishment of new industry to 85% for clearing derelict land and waste tips. Successive attempts have been made to bring the National Coal Board's financial structure into line with economic reality by writing down the value of its fixed capital assets: in 1965 the Government wrote off some £400m. of the N.C.B.'s capital debt to the Exchequer which reduced the interest payments in 1965–66 by an annual £30m. Since 1961 the Treasury has set a target for the N.C.B. that it should make a surplus of £10m. a year towards replacement of fixed assets over and above depreciation of historic costs. This has recently been revised to a requirement to break even. The estimated value of the 1966–71 capital reconstruction programme was £380m., but despite the general rise in coal prices in 1966, resulting in a surplus of £0.3m., no contribution was made for the extra depreciation.

In the 1967 Fuel Policy Review the prognostications for coal foresaw a continuing decline in coal production until at least 1975, when internal demand was estimated to be only 118m. tons. By 1970 it was already down to 155m. tons. The change of Government in 1970 saw a change in the details of the coal protection policy, although circumstances had changed owing to a shortage of particular grades of coal. Open importation of coal was permitted and more conversion of stations to dual firing was allowed. The grants and subsidies to redundant miners were continued, but changes in regional development policies were instituted which will reduce the net subsidy to the declining coalfield areas.

A comparison in cost-benefit terms has been published by an economist[8] employed by the U.K. Atomic Energy Authority which attempts to strike a comprehensive economic balance between a nuclear fuelled station and a coal fired station. This paper is complemented by a later attempt[9] which included, in the cost benefit equation for choice between coal firing and nuclear fuel, a calculation of miners' redundancy cost. The earlier paper is more comprehensive in attempting to include, in addition to the purely economic aspects of the comparison considerations as to the environmental aspects. Under this heading are included the relative costs of illness and injuries in employment, the safety of the general public, the costs of air pollution and the dumping of wastes. There appears in this area of policy, at least, to be no formal mechanism by which such attempted quantifications of amenity costs are brought into direct confrontation with the financial quantifications, though they are clearly used as a means of influencing informed public opinion.

Thus, while the use of such cost benefit analyses is becoming more

8. M. Phillips, "A Broader Approach to Benefits from Nuclear Power and Associated Social and Other Costs," *ATOM*, Vol. 145 (November, 1968), 297–307.

9. G. K. Jones, *CBA of the Choice Between Nuclear and Other Fuels*, Paper presented to the O.R. Society Conference, Eastbourne, U.K., November, 1969.

widespread in official decision making and while the methods do not lack their advocates,[10] it is not certain whether their closely documented reasoning has ever had more than a marginal influence upon the final decision in the energy policy field. In any other area—except perhaps transport, where the social cost-benefit approach justified the Victoria underground line—the situation is not much different, as witnesses the fact that the recommendations of the largest example of the method, the Roskill Commission Report on the Third London Airport, have been set aside.

As a hindsight exercise a limited study has been made of the socio-economic implications of siting and constructing a 500MW nuclear power station and a 320MW capacity pumped storage scheme in the thinly populated area of North Wales, an area formerly heavily dependent on the slate industry and agriculture but latterly declining. The construction work at the station reversed a steady population decline and a one per cent annual increase was substituted. Within four months 491 local people (84% of work force) were employed, rising to over 1500 ($\sim 50\%$ of work force) and the wages bill', spent locally, was annually £3m. Commercial life, was thus stimulated for up to 40 miles from the sites. In terms of lasting effects, of the 360 permanent station staff 64% were recruited locally, and major road and port improvements attracted increased tourist and other trade. As the availability of suitable sites decreases, second order social and economic considerations such as can be derived from the above example will be used increasingly as arguments in siting policy by the Electricity authorities, though at present they do not constitute a very major consideration.

Forecasting of Power Demands

Economic and social considerations are, of course, taken into account by forecasters within the electricity industry[11] and attempts are made by the Generating Board in collaboration with the Electricity Council to relate such items as historical increases in Gross Domestic Product and industrial electricity consumption.

In the period 1922–62, four distinct periods with different growth rates have been distinguished in which increases in production and electrical consumption can be related. Such correlations leave unexplained some percentage of consumption increases which are usually explained in terms of:

10. T. C. Sinclair, "The Incorporation of Health and Welfare Risks into Technological Forecasting," *Research Policy*, Vol. 1, No. 1 (Amsterdam: North Holland Publishing Company, November, 1971); and R. McKean, *Efficiency in Government Through Systems Analysis* (New York: Wiley and Son, 1958).

11. G. England and P. A. Lingard, *The Electricity Supply Industry and Economic Growth*, A paper presented to the British Electrical Power Convention, 1964.

a) Substitution for coal.

b) Improved industrial labour productivity—rate of mechanisation.

c) Technological progress—which will include estimates of the probable rate of advance of anti-pollution measures and their cost.

Thus, forward estimating in these three fields must be attempted by the electrical power industry.

The growth of commercial usage of electricity, though relatively small, has shown a larger percentage of annual increases than industrial and domestic consumption. The great diversity of commercial users makes analysis of the causes of growth difficult and forecasts more uncertain. Some extrapolation is possible, however, in the case of the provisions for increased amenity and lighting standards made in the recent Office and Shops Act which will be one of the social factors accounting for growth in this sector.

In relation to domestic usage, trends are forecast in relation to consumer ownership of central heating and electrical appliances since it is the increase in consumption per head, rather than in numbers of consumers which has caused the bulk of the increase in this sector since the 1939–45 war. The extension of electrical supply to all consumers is a statutory duty and an occasional source of financial loss to the industry.

The nationalised electricity boards are concerned in the main with the provision of "an efficient, co-ordinated and economical system of electrical supply." However, social obligations placed on the Boards affect their operation as commercial organisations. Area Distribution Boards in England and Wales in 1966 were still losing £4–5m. p.a. on rural electrification which amounted to 14% of the total rural revenue. By the late sixties over 95% of farms had been connected up to the supply. In Scotland the North of Scotland Hydro Board has a duty "to collaborate in carrying out any measure for *social* improvement and economic development in the North of Scotland."

Environmental Pollution as a Factor in Policy Making

One aspect of increasing importance is the concern of the power industries with the social implications of their activities, e.g., that of disamenity and environmental pollution problems. Recently the First Report of the Standing Commission on Environmental Pollution[12] showed that in the fifteen years from 1953 to 1968 the relative amounts of smoke emitted by industry (including electrical power), railways, and domestic sources had radically altered. In 1968, 80% of all smoke came from domestic sources against only a little more than 50% in 1953, and smoke concentrations had

12. Sir Eric Ashby (Chairman), *Royal Commission on Environmental Pollution*, Command Paper No. 4585 (London: Her Majesty's Stationery Office, 1971).

fallen from around 175 microgrammes per cubic metre to around 60. Sulphur dioxide emissions had risen from 5.0 million metric tons total to almost 6.0 with electricity works doubling their contribution, though ground level concentrations had fallen from 175 microgrammes per cubic metre to around 110. These figures have been achieved against a 10% population increase and a 1.7% annual gross energy consumption increase.

The Generating Board's position in the light of the foregoing is that it feels fairly secure against criticism. The pollution of air and water and from noise or land dereliction do arise as local problems, but the Board considers that the bulk of its research and development work has been done and has solved the basic problems in this area. Pollution control measures account for £18m. p.a. expenditure (of this £17m. is on air pollution). Lessening expenditures are forecast since coal will become a smaller percentage of fuel consumed and alternative fuels, particularly nuclear, are used. The Act reconstituting the C.E.G.B. placed on it the duty, in planning, of taking account of potential damage to flora, fauna, local monuments, etc. and now all industrial development is covered by Town and Country Planning Acts. The Board claims, however, that the impetus for most of its anti-pollution programmes are derived from its desire to present the best possible public image and from its concern to preserve and enhance amenity.

The planning constraints that pollution and amenity considerations place upon the Generating Board appear principally as time delays in starting building operations if Ministerial consent is withheld until a public inquiry has taken place. The care taken by the Board in selecting sites and sounding out local opinion on the pollution and visual amenity aspect is reflected by the fact that very few public inquiries have been held—two in the early 1960's, one in 1970, and two are scheduled for this year—and that only one public inquiry resulted in a refusal of Ministerial consent for a major power station.

It is important to consider the background of the amenity and pollution problem within which the power industry in Britain is working. Two points are important. The first is the great shortage in Britain of inland water resources suitable for power station cooling, which in effect severely limits the number of possible inland sites for the large stations required by the British power system. Transport considerations in the case of coal and safety considerations in the case of nuclear power have dictated the Board's policy of concentrating much base load coal fired capacity in the Midlands near relatively cheap coal supplies and particularly along one large river (the River Trent), and of concentrating nuclear base load capacity in remote coastal sites. In the case of coal the long-term employment benefits to the local coal mining labour force have been obvious and have therefore muted local objection to building coal fired stations. The remote siting of nuclear stations in sparsely populated areas has meant that there has been no mass public complaint, although objections of small groups and individual naturalists and others interested in visual amenity have been vociferous. The second important point is that for reasons of government economic policy the number of large base-load oil fired stations in Britain is very small indeed, so that, unlike Japan for example, the overall problem of sulphur dioxide emission has been correspondingly smaller than if a high proportion of the base-load thermal plants constructed in the last decade had been oil fired. Had the latter been the case the problem of SO_2 pollution would have been much more serious and it is open to question whether the Board's policy of relying solely on atmospheric dispersion would have sufficed.

The current resort to public inquiry is probably related to the mounting current concern with visual amenity and a generalised sensitivity about pollution rather than a greater real destructive threat associated with the siting of power stations. An interesting footnote here is that C.E.G.B. planners feel that individual objectors are now more numerous and vociferous than ten years ago when most public complaint came from organised groups. One result has been that inquiry proceedings have increased from 3-6 months to 6-9 months. This lengthening must, however, be seen in context, for it has had a far less delaying effect on construction programmes than delays from the withholding of government approval for financing those programmes, arising from the state of demand.

In as far as it is possible to produce an inventory of U.K. power stations on an anti-pollution technology basis, the large building programme of the sixties has meant that over 50% of C.E.G.B. capacity has the latest anti-pollution devices and most of the rest is "reasonably adequate" to reach the standards of the Alkali Acts. During 1963–66 a programme of improving the pollution characteristics of old stations was carried out, costing £15m. Occasional local pressure about existing stations is experienced from, for example, the Member of Parliament in the neighbourhood, to upgrade a station's anti-pollution technology.

The technical choice of station characteristics, fuel etc. is seldom decided on purely pollution grounds, except where plant size or regional capacity may have an upper limit placed on it because of the local characteristics. In one site recently the stack height was limited by the proximity of an airport. In turn this caused the station capacity to be reduced. Other probable, though as yet untested, possibilities are that existing density of power station or local topography in the shape of nearby hills would act to limit output because of pollution possibilities.

The general strategy adopted in the U.K. by the Generating Board to limit ground level concentrations of pollutants may be summarised as follows:— smoke is eliminated or reduced at emission, dust is eliminated or reduced both by filtration at the point of emission and by dispersing it at a height, and sulphur dioxide is dispersed from high stacks.

Research Policy

Environmental research and development presently amounts to some 6–7% of the C.E.G.B.'s total R. and D. budget of £10m. and in percentage terms this is larger than the average value for most U.K. enterprises carrying out widely based R. and D. programmes. This proportion shows some signs of decreasing slightly in the next few years as the bulk of the air pollution research is completed. The expenditure on air pollution R. and D. fell from £182,000 in 1968 to £85,000 in 1970. The 6–7% outlay in this year covers problems in the following fields: air, water and noise pollution and biological research. A breakdown for 1966–67 for expenditure in the nine laboratories of the Board shows a similar percentage devoted to pollution R. and D.:

Nuclear power	21%
Conventional fuel	20%
Electricity generating equipment	5%
Environmental problems	7%
Transmission and distribution	14%
Uses of electricity	11%
Special projects	13%
Miscellaneous	9%

Until 1958 the supply industry had to rely upon the manufacturers and the development of an R. and D. programme starts from that year. Joint research programmes are pursued with manufacturers on a 50–50 basis, the current annual value of the effort here being £2m. R. and D. on anti-pollution devices has largely been initiated and materially supported by the C.E.G.B. They have carried out well over half the basic physics research on filtration and air dispersion problems, though considerable engineering development has been placed in the hands of the manufacturers. Collaboration has been maintained through a joint panel. The Board's practice is to fully detail specifications for the technical plant and this has allowed it to bring considerable pressure in this field too. Since the general achievement of a 99.3% particle arrest level in filtration technology in 1955 the pressure for R. and D. work in this area from the Board has been towards the achievement of greater reliability with the development work being done largely by the Board. In this area the major incentive for manufacturers is the possibility of a large export market. The situation with regard to R. and D. on sulphur removal is that the Board is dissatisfied with the economics of currently available methods and would attempt such development in their own laboratories if they genuinely believed any other potential method could compete with the present solution of high-velocity dispersal from a stack.

An area where strong public opinion, as expressed, for example, during public inquiries, has initiated and sustained a large research programme within the C.E.G.B. where many engineers entertained considerable scepticism, is that of the undergrounding of transmission lines. After several years of research the costs per mile of undergrounding lines is still calculated by C.E.G.B. engineers at about 17–18 times that of overhead lines. The achievement of success in this area appears to the Board as being unlikely and a considerable environmental effect would still be caused even if such methods were used. Transmission lines generally give rise to greater public outcry over disamenity, especially in areas of high visual amenity. Hence if the ratio of costs of overhead transmission to underground transmission became more reasonable it is probable that the Board would come under considerable pressure to apply a high social cost factor in their calculation in favour of underground transmission.

Conclusion

Some of the criteria and factors which form the bases of U.K. energy policy with regard to electrical power production have been reviewed. The legislative and institutional framework within which these considerations apply have been described particularly in respect of the social effects including amenity and pollution which have to be taken into account. Energy policy is of fairly recent genesis and has in its short life been subject to considerable vagaries partly because of the incompatibility of policy objectives. The present picture is one of, at a political level, an attempt to reduce official intervention while taking into account necessary social obligations; at the bureaucratic level, to increase flexibility and to keep open options; at the technical level, to encompass a fairly rapidly altering technological position. Underlying and modifying all these postures are considerations relating to possible environmental effects often arising one stage removed from the actual site of power production. It seems likely that energy policy will remain in continuous evolution for some time, with a move towards more explicit evaluation and account being taken of environmental considerations, though the trend is at present slow. This arises mainly from the relatively satisfactory state of present pollution control and favourable historical development. It remains to be seen whether increased social concern will produce any mayor policy changes.

Table 1. Industries' estimates of fuel demand in the United Kingdom resulting from the Industrial Inquiry for the National Plan.

	Million tons of coal equivalent[a]				
	1960 (Actual)	%	1964 (Actual)	%	1970 (Estimate)
Coal:					
For power stations	51.9	26.4	68.0	36.4	84
For gas works	22.6		20.5		10
For other purposes	122.2		98.7		81
Total	196.7		187.2		175[b]
Oil (including petroleum gases):					
For power stations	9.2		9.7		14
For gas works	1.9		5.0		14½
For other purposes	54.4		78.6		115½
Total	65.5		93.3		144
Natural gas	0.1		0.3		1½
Nuclear power and hydro-electricity	2.6		5.1		16½
Total inland demand for energy	264.9		285.9		337
Electricity (thousand million kilowatt hours)	104.9		143.4		241
Gas (million therms)	2,636		3,014		4,635

a) Except for electricity and gas.
b) It is estimated, on present prospects, that a further 5 million tons may be exported, giving a total estimated demand for coal of 180 million tons. In 1960, coal exports were 5.5 million tons and in 1964, 6.0 million tons.

Table 2. Ministry of Power's Sector Analysis of Inland Energy Demand

	Million tons coal equivalent[a]		
	1960 (Actual)	1964 (Actual)	1970 (Estimate)
Iron and steel	34.8	36.1	37
General industry	77.2	83.8	98
Railways	11.0	7.2	4½
Other transport	22.5	29.1	38½
Domestic	71.5	78.7	86
Other inland	46.7	50.4	59
Total	263.7	285.3	323
Adjustments for stock changes and exports of coke and manufactured fuel	1.2	0.6	1
Total inland demand for energy	264.9	285.9	324
Adjusted to normal temperature	265.4	285.4	

a) Secondary fuels (electricity, gas and coke) are equated to the quantity of primary fuel (coal, oil and natural gas) required to produce them. Thus the table indicates primary fuel input and the final total can be compared directly with that of Table 1. In arriving at the demand in 1970 for each sector shown above, allowance has been made for the different efficiencies of use of each fuel.

Table 3. Projections of long-term growth in U.K. energy consumption.

Year	Energy co-efficient of 0.7 0.6 (from 1955 base year)		Extrapolations of	
			Plan forecast from 1970 at same rate of growth in energy use	NIESR forecast[a] from 1975 at same rate of growth in energy use
1970	342	326	324	(342)
1975	377	354	379	390*
1980	415	385	421	445
1985	458	418	467	508

a) Midpoint of forecast range.

Table 4. Primary Fuel Use in 1970.

	1966 (Actual)	1970 (Forecast)	1970 (Actual)
	Million tons coal equivalent		
Coal	174.7	152[a]	154.4
Oil	111.7	125	145.6
Nuclear and Hydro-Electricity	10.2	16	11.9
Natural Gas	1.1	17	15.3
Total Inland Demand for Energy	297.7	310	327.2

a) Plus an estimated 3 million tons for exports, making a total demand of 155 million tons.

Chapter VII

THE FRENCH POLICY IN THE ELECTRICAL ENERGY AND
ENVIRONMENTAL FIELDS

by *J. M. Martin*

France, like all countries having reached a certain development level is
confronted with the problem of the mastery of technical progress under a
double aspect: to limit the harmful effects, whether foreseeable or not, on
the whole society without interrupting the rhythm which commands economic growth; and to orientate future progress, even provoke new progress,
in order to satisfy the real needs of man and society. These preoccupations
are beginning to radically modify the social science concept of apprehending
reality (new concepts of needs and of the relationships of man, society, etc.)
and of their methods of doing so (multidisciplinary, prospective studies,
research on social indicators, etc.). The concrete introduction of these preoccupations is only achieved slowly inside public policies which are supposed
to put aside individual interests for the benefit of collective interests. Electricity policy is not an exception to this general statement. If forecasts for
the consumption, production and the technical characteristics of production
means are made today for the end of this century, one is ill equipped to envisage
the effects of expected growth on the rest of the economy, on transport, on
urbanism, on the rural surroundings, on international relations. Furthermore, the methods for constructing this electric power growth in relation to
a certain picture of the wished-for society, both domestically and internationally, would probably be considered as utopic. The question can,
however be asked as to whether technicians and social science specialists
should not work together in this direction if one believes that progress is
primarily the mastery of technical progress. The following paper will not
touch on such an ambitious question. We shall simply try to analyse what was
the French policy in the electricity field and the problems involved. We thus
distinguish the following points:
– the situation of the French electricity industry and its perspectives,
– the electricity industry within the energy policy,
– the electricity industry and the environment,
– a few lessons for the future

The Situation of the French Electricity Industry and its Perspectives

The electricity industry in France presents certain characteristics both from the total growth and participation in the satisfaction of energy needs point of view and from the internal, technical, economic and institutional structure point of view.
The total consumption (losses included) of 130,7 tWh in 1969 is to be placed within a growth marked by a clear inflexion around 1964/65. Having up to then feen doubling every ten years, i.e., an annual average rate of 7,2%, this growth fell to 6,25% during the last five years. This growth rate fall, which was essentially due to a slow-down of the consumption evolution of industry (especially energy intensive industries), accentuated the fact that France was already behind in this field in the domestic per family electricity consumption. Hence, the electricity participation in final energy consumption did not reach the Vth Plan forecasts and remains lower than observable rates for most highly industrialised nations. The following table shows this phenomenon:

Table 1. Proportion of Electricity in final energy consumption in 1968[a])

	All sectors	Industry	Residential tertiary
Japan	29,9%	43%	28%
North America	25,2	39	32
Western Europe	25,1	35	27
Federal Germany	22,7	32	23
United Kingdom	26,1	30	34
France	21,6	32,6	19,9

a) Cf. General Plan Commissariat, *Report of the Energy Commission of the VIth Plan* (Paris, March 1971), p. 59. Let us recall that these forecasts are only approximate because of conversion difficulties of electricity into common energy units. We retain here 1000 kWh = 1/3 TCE.

This unforeseen slow-down, which to date remains insufficiently explained, incited the authors of the VIth Plan to foresee future growth more prudently. In spite of their marked preference for the relaunching of a very high electricity consumption growth (8% p.a. at the end of the VIth Plan period), they retained two hypotheses as follows:

Table 2 : Total electricity consumption forecasts (including losses) in tWh.

	1975	1980	1985
Upper hypothesis (success of plan objectives)	207	295	420
Lower hypothesis (semi-failure and growth rate staying at 6,5%)	192	265	360

In both cases, electricity will probably not reach more than 30% of the final energy consumption in 1985, whereas the Vth Plan envisaged 43–44%.[1] Even on this basis, however, the gap between both hypotheses is large and leaves an important choice margin for an electricity policy, especially when 50% of final consumption could be reached by electricity at the end of the century. This is even more striking if one analyses the branch structure i.e. the proportion of different electricity production techniques, the respective importance of the different types of fuels, the localisation of power stations and the transport/distribution network, etc.[2] In the production field, the only one to be examined here, France's evolution was rather similar to that observed in other industrialised countries. Several specific characteristics should, however, be pointed out.

Table 3 : Past and foreseen production structure in tWh[a])

	1960	1970	1975	1985
Total consumption	72	140	200	400
Hydroelectric production	33	52	57	60
Non-EDF thermal production	19	25	28	34
EDF thermal and nuclear production	20	63	115	306

a) Cf. Jean FERONS, "Fuel supplies to EDF Power Stations," R.F.E. special n[0] op. cit., p. 241.

1. Cf. General Report of the Energy Commission, Vth Plan (1966/70), Imprimerie Nationale, p. 152.
2. A good synthesis on all these points can be found in the special number of the R.F.E.: (French Energy Review) on the 25th anniversary of Electricity and Gas of France, March/April 1971.

The preceding table shows that from the early 60's hydroelectric production began to lose the important place it held beforehand (approx 50% of the production) and that its future growth will be insignificant. The "qualitative" importance which pumping could provide to meet peak demands should not be forgotten. Although apparently condemned to the same destiny, non-EDF thermal production could, however, give some surprises on a long term basis if autoproduction inside vast industrial complexes was to be developped. Finally, the EDF thermal and nuclear production which is supposed to ensure most of the expected growth of electrical domain necessitates a certain number of choices between production techniques and fuel supplies. With a nuclear-based electricity production of 4,5 tWh in 1970, France (whose nuclear programme was seriously slowed down during the Vth Plan) is now lagging behind other industrial countries. With a conjuncture marked by the abandonning of natural uranium type and a reversal of fuel price trends, the VIth Plan chose a nuclear programme of 8000 MW which should result in a production of 56,5 tWh in 1980, i.e., 20% approximately of the electricity produced at that time. Such a choice—if it is confirmed—would mean France would build as many nuclear-based as *conventional* thermal-based plants during the VIth Plan, then solely nuclear-based plants during the 7th Plan in such a way as to satisfy 50% of the non-peak needs in 1985 through this new source. The rest of the thermal production would be essentially fuel-based although the VIth Plan has recommended continuing "possibilities to supply the greater part of thermal power stations which should be converted to fuel fired with coal, and when the international coal market is more easy, the partial supplying with coal of some of them."[3]

What are the natures of the policies which result in the choices leading to the present situation of the French electric industry? What are the objectives and criteria used? What is their respective weight for the different decision centres directly or indirectly associated with this policy elaboration? What are the factors which explain the different observable combinations of objectives and their temporal evolution?

The examination of the French energy policy can give some elements of reply to these questions.

The Electric Industry in the Energy Policy

In countries where the electric industry is partly or entirely in private hands, the often somewhat arbitrary distinction between firm strategy public policies offers a great facility. The French situation resulting from post-war

---

3. *Report of the Energy Commission of the VIth Plan, op. cit.,* p. 6.

nationalisation laws does not allow us to use this facility for reasons appertaining more to fact than to law. In fact, nothing seems to prohibit EDF, in the laws, and within the limits imposed upon a public service, of using the same set of management rules as private enterprise. We recall here that since 1946, EDF has the electricity transportation and distribution monopoly. EDF also ensures 72% of the production, the remainder coming either from private industrial production (10%), or from the SNCF and the CNR for hydroelectricity (9%), or from the French Coal Board for mining-based production (9%). In the following pages we shall assimilate EDF and the electric industry because of the former's leading role towards the remainder (excluding private production).

EDF is a national public corporation of an industrial and commercial nature, and according to Article 4 of the Nationalisation Law, "follows the usual rules of industrial companies for its financial and accountancy management." It is therefore not subjected to the State Markets Code and is only subjected to control "a priori" of its contracts by a commission working under the auspices of the Central Markets/contracts Commission. This control is extremely supple and never (or rarely) seems to have questioned a contract made by EDF. It would, however, be erroneous to deduce that the electric corporation is free in all movements. In fact, since its creation, it has been shadowed by the public authorities and it is often difficult to detect in its policies (industrial, scientific, commercial, financial) that part, which is decided by the corporation and the part imposed by the government. The latter's control in the broad sense, can be presented in the following way.[4]

– The total volume of annual investments and their calender for the year rely on Finance Ministry permits for the financial credit from FDES transit through the national budget. There is no doubt that this aspect of control is particularly efficacious, for it commands the investment programmes and hence the firms' growth rate and its effects on the capital goods industries.
– Furthermore, this same Ministry has always closely controlled both the tariff evolution and the employees remunerations. On this latter subject, the recent progress contracts signed with EDF, as with other state firms, are direct effects of the incomes policy decided by the state authorities.
– Strictly speaking it is the Industry Ministry which exercises the state control on the electricity corporation. The Ministry's Gas and Electricity Direction is most directly concerned with following the electricity industry's activities to ascertain whether these activities are in conformity with the laws, decrees, rules and general orientations of the energy policy. However,

4. These problems are analysed in more detail in an unpublished paper: E. ROTH and J. M. MARTIN, "The International Relations in Innovation Genesis," (Grenoble 1968), 320 p.

other directions of this same ministry also intervene every time an EDF decision has repercussions on other sectors. This is the case, for example, of the Mechanical and Electrical Industries Direction (DIME) which has always surveyed with precaution the public sectors capital goods orders, as well as for the Mining Direction and the Carburants Direction which have always kept a close and interested eye on the fuel supply choices of thermal power stations. Since 1965 the action of the three Directions on energy problems has been coordinated by the Energy Secretariat but it would seem that this latter's action was not imposed without difficulty.

It is thus through these different administrative channels that the electricity industry is most directly subordinated to the Energy policy which covers the electricity branch as well as the fuel branch. In order to be relatively thorough, we must, however, also mention the less formal but more numerous procedures which associate EDF representatives with works of administrative organisations such as the DGRST for scientific and technical research, the DATAR for site planning, the Superior Water Council, etc.

What is this energy policy? By whom and how is it elaborated? What are the objectives and how are they materialised in the choices of the electricity industry which we briefly reviewed above? Normally it is with the preparation of each Plan that the energy policy is elaborated.

It is useless to recall that these plans are only indicative, both for the government and the public sector although the latter has to outline objectives and not just forecast.

This, however, did not prevent frequent new analyses of investment programmings, (slowing down or speeding up of programme permits according to the conjunctural economic policy), even in important economic and technological choices (i.e. modification of nuclear reactor types during the Vth Plan). During the past 25 years the central objective in the energy field, and particularly for electricity, has been to satisfy the nations needs, but the criteria used to attain them varied. Without entering into a very detailed analysis, let us briefly recall:

– that the 1st Plan's objectives were to rapidly open the existing installations, to construct new ones using the national energy potential (above all, hydraulic), while rendering more rational and reorganising the networks inherited from the old electricity companies (interconnection);

– that during the '50s, the electricity industry was still mainly concerned with the development of the production means but that the choices began to be marked by capital cost and foreign currency cost of electricity (long or short term preference, preference for a more costly national energy or a foreign energy more costly in foreign currency)...;

– that during the '60s, the arrival of nuclear energy, the favorable petroleum conjuncture and above all the frontier barriers levied, all contributed to new possibilities and rendered choices more complex, entailing a more deep reflexion on the energy policy.

The 4th and 5th plans (1962–65, 1966–70) reflect this new orientation and its consequences on the electricity industry. Apart from a few differences, the main objective is the same:

"The objective of an energy policy should be to satisfy the user's needs, at the lowest cost for the nation, whilst respecting certain imperative conditions, of a national or social nature, such as supply security or regional activity equilibrium which, in any case wherever possible should also be expressed on a cost basis."[5] The definition adopted by the IVth Plan was both less explicit and more ambiguous:

"The long term energy policy should aim to satisfy needs at lowest costs, i.e. supply power to users at the lowest price."[6] It was, however, accompanied by a more detailed analysis than that of the Vth Plan in so far as the limits of this objective and the nature of other objectives were concerned.

This definition calls for two groups of comments.

It explicitly refers to an economic pattern (organisation of the economy and rationality of choices) in terms of competition, of free consumer and producer initiative, of production factors and allocation by prices which reflect true costs. But what are these costs? Planners admit that these costs are not those only borne by the firms alone but that they should include those borne by the whole collectivity or imposed on the firms. We refer particularly to the conversion costs of miners, hit by the progressive shutdowns of the collieries, to the costs of security stocks (3 months) for petrol, to those affected by pollution reduction measures, etc. If, as the Vth Plan authors wish, all the effects of electricity growth on the remainder of the national economy could be reduced to external economics which could be measured and integrated into costs, the application of such a policy would be simplified. The prices and costs which reflect the real expenditures paid by the collectivity, would constitute excellent choice indicators; any other intervention of a regulatory nature or pertaining to quantities would be avoided, the public authorities' job being limited to guiding the conditions of a healthy competition (respect of general legislation, public and non-discriminatory tariffs, etc.).

The facts, however, are somewhat different. Including collectivity charges in the cost is, first, neither always possible nor sufficient. Thus during the 60's, for example, EDF provided its thermal power stations with domestic

5. *Report of the Energy Commission of the Vth Plan, op. cit.*, p. 17.
6. *Report of the Energy Commission of the IVth Plan, op. cit.*, p. 17.

French and Sarre coal which was far more costly than imported coal or fuel. Although this was the result of an EDF-CDF contract, this energy supply scheme was a direct part of the governments' coal policy.[7]

Second, some objectives of the public authorities cannot be precisely evaluated and are often given as outer limits not to be transgressed (for example, in the geographic concentration of petroleum supplies), as imposed or suggested preferences (for domestic electrical equipment), as technologies upheld or encouraged because they satisfy a certain conception of national independence (natural uranium nuclear power stations), as imposed tariff modifications so as not to penalise less developed regions through more costly electricity (the famous Brittany franc!). From these few examples, it is clear that the electricity industry's choices within the French energy policy have not exactly respected the competitive pattern of reference. But can this be called irrational? Certainly this is irrational if one aligns to a strict efficiency rationality, but is this the only conceivable rationality for choices which interest either the whole collectivity or broad sectors of the economy and the society? This is not obvious and we shall return to this debate.

If the French electricity industry makes its main choices within the more or less constraining framework of an energy policy it would be wrong to deduce that it has simply submitted itself to this policy and passively adapted itself. We must recall that EDF actively participates in the elaboration of the entire energy policy, especially in its electricity part. Although both economic and sociological analyses concerning its influence on main orientations are lacking, it would seem that this influence has never been negligible. Andrew Shonfield writes, "from several standpoints, the development of French planning in the 50's can be considered as a voluntary collusion between top civil servants and managers of large firms."[8] Amongst the latter the EDF managers were certainly not the least listened to. Yet, until recently, EDF's policy was never that of a firm only worrying about its sales and profits. Although its main work was directed towards a per kWh price reduction, it simultaneously fulfilled certain missions which in a liberal economic logic could seem as belonging more to public authorities than to the firm even if this latter is public-owned. These are obviously suggestive appreciations, for it is most difficult to mark out a precise frontier between public authorities and firm functions. A few examples, nevertheless, could clarify this problem.[9]

7. A 14 year contract from 11.1.65 referring to maximum deliveries of 15 Mt in 1978, Jean Feron, "The Supply...," *op. cit.*, p. 242.
8. Andrew Shonfield, *Le Capitalisme Aujourd'hui*, (Gallimard, 1967), p. 132.
9. Roth and Martin, *op. cit.*

– EDF, as an important purchaser of capital goods, had progressively elaborated in practice a policy (in the true sense of the word) to be applied to the domestic heavy material and electrical construction industry. By reserving its orders to this industry, by encouraging the firms to specialise and to amalgamate until they reached the proposed optimum number of approximately three firms per branch, by imposing often original norms and specifications, the electricity corporation anticipated certain aspects of the industrial policy to be applied by the public authorities after 1965, in a context which was in fact very different.[10]

– At the same time EDF played a particularly active role in research and application of new technologies. This activity could be considered as normal for any innovating firm, if it had not been accompanied by a special diffusion effort of the results, of promotion of new techniques for the constructors; all these aspects anticipated actions which were undertaken later, in other fields, with the DGRST.

Is EDF's behaviour identical as far as environmental quality and nuisance suppression are concerned? We now analyse this aspect.

The Electricity Industry and the Environment

The French energy policy, preoccupied by supply security, regional equilibrium, and national independence, as can be ascertained in the 4th, 5th and 6th plans, is virtually silent on environmental problems. They are hardly mentioned, except here and there in order to show that the problem exists but there is no analysis of its consequences or of the particular aims which should be envisaged. How is the problem at present discussed, and what are the respective roles of the public authorities and electricity industry?

The synthesis written on this subject at the most recent UNIPEDE congress can be a starting point.[11]

10. Naturally, this description is so rapid that it verges on a caricature. The Public authorities were already partly responsible for this behaviour which did not equally please all the directors of the corporation. We wish only to underline the difficulty of marking the limits of a public policy which could easily be seen as having an opposite nature to the strategy of a firm.

11. F. Faux, "Environmental Problems Created by Electricity Production," II.3., Cannes, September 1970. Table 4 is taken from this document.

Table 4 : Nature of obligations concerning the exploitation and construction of thermal power stations in France.

	Legal obligations	Exterior benevolent actions	Spontaneous actions of the electrical industry
General protection of the surroundings	x		
Visual protection			x
Protection of existing surroundings			
historic buildings and sites	x		
regions of natural beauty	x		
flora and fauna		x	
Perturbation			
noise			x
radio and television			x
Effluent control			
solids	x	x	
liquids		x	x
gaseous	x	x	
radio-active	x		

To date, general French regulation (this referring to all types of industrial installations) covers four environmental aspects:
– site protection, i.e. regions of natural beauty;
– protection of historic buildings and monuments;
– water protection;
– general atmospheric pollution (law on classified establishments).
Up to the very recent creation of the Environment Ministry, these regulations were applied by several ministries and organisations (cultural affairs, social affairs, agriculture, DGRST, etc.). This dispersion could only weaken the influence of an already old law, although this latter was partially renewed every so often, and which in any case could not be applied without the active participation of the electricity industry. But, here again, in this

field, EDF did not ignore its responsibilities concerning pollution as is shown by proven results and by presently applied aims.

Pollution can be particularly serious in the electricity industry because of cooling-water and gaseous rejects (NO_2, SO_2, dusts) from thermal power stations. Nuclear power stations, do not produce smoke but have a more accentuated thermal action on the rivers and present radio-activity and tritium escaping risks.[12]

In so far as atmospheric pollution is concerned, the French legislation at the beginning of the '60s was based on an old law of 19 December 1917 on unhealthy establishments and by the law of 20 April 1932, Morizet law, which covered all the industrial smokes.[13]

"The imprecision of this law had however contributed to making execution measures which could be taken inefficacious. Its influence was even restricted by the insufficient number of inspectors and agents to control its execution."[14] These defects were at the reason of the new law of 19 December 1960 which sets more precise norms and foresees sanctions in the administrative, legal and civil fields. But, already at that time, EDF had taken a series of actions to limit atmospheric pollution by installing dust-reducers in all its power stations (with a 90% working rate in 1959), and by further raising the chimney stacks and by using low-sulphur fuels.[15] The following table shows the good position of EDF dust reduction efforts in 1963.

Table 5 : Origin of atmospheric pollution in France in 1963 (dust rejected into the atmosphere in Kt and %)[a]

Cokeries	140	12,5%
Medium and large industry	440	36,6
Small industry	190	15,6
Steam locomotives	100	8,2
House hearths	25	2,1
Thermal mining power stations	255	21,6
EDF thermal power stations	40	3,4
Total	1190	100,0

a) R. GINOCCHIO, "EDF Participation in the Anti-atmospheric Pollution Struggle," *R.F.E.*, (December 1964), p. 117.

12. Radio-activity risks were even higher in the installations for the preparation and the re-treating of fuels.

13. This was indirect, working through Prefectoral measures, as the law itself was very general.

14. M. Potier, *Energy and Security* (Mouton, 1969), p. 155.

15. Here, EDF had the natural advantage by using French coal with a low sulphur content.

EDF's efforts can be measured more accurately if one remembers that its power stations contributed 8 % of total pollution by dust in 1958 and this is quite low when one compares this percentage with that of mining power-stations whose installed capacity is less. Although already expensive, both from the investment and running-costs point of view,[16] the methods to combat pollution have been permanently developed since then and in several directions.[17]

– The reduction of the quantities of polluting material by the inventing of a high capacity electrostatic dust reducer for coal-based power stations, by the use of BTS fuel in the others only in case of unfavourable weather conditions, by smoke neutralisation through gaseous ammonia injections, by better dispersion thanks to raised chimnies (up to 240 m.).

– At the same time methods for surveying the pollution of conventional power station sites have been reinforced by the systematic installation of SS machines (SS: sulfursmoke) whose most recent models were conceived by EDF research. Air and biological controls are naturally permanently undertaken around nuclear power stations.

– The research programmes undertaken by EDF concern both the reduction of polluting agents (principally SO_2), the scientific study of dispersion of combustion gas mixtures and nitrogen oxides (NO_2). In so far as SO_2 is concerned, the EDF research work is concentrating on desulphurisation of combustion gases in a 25 MW pilot unit, because of the absence of a low-priced desulphurised fuel (at present the increase can be evaluated to be between 10% and 20% of the price of the fuel for a sulphur content lower than 1%). Desulphurisation without recuperation could be put under way towards 1975 on 250 and 600 MW slices if an approximate 12% price rise of the Kwh was applied. This cost could be considerably reduced if sulphur and its derivatives could be recuperated.

In the water pollution field, French legislation has been belatedly adapted, for before the 16th December 1964 law on "the regime and repartition of waters and anti-pollution campaign", the only legislative texts applicable were more than twenty years old.[18] Generally speaking, rules were fixed and

16. M. POTIER, *op. cit.*, p. 152, evaluated them as follows:
– investment costs: 7 000 000 F for a 250 MW slice i.e., 4% of total investment; this sum includes the dust-reducer (5 000 000), the stocking installations for flying cinders (1 500 000), the raising of the chimney,
– running costs: 200 000 F/year for the same slice, i.e. 3% of running costs, fuel not included, plus the energy consumed by the dust reducer (0,5% of the energy produced) and the supplementary cost of non sulphurous fuels.
17. B. de RETZ, "EDF's Action to Prevent and Survey Atmospheric Pollution," *Annales des Mines* (November 1970), pp. 31–44.
18. Rural code and law of 19 December 1917, modified on 20 April 1932 and 21 November 1942. (continued on p. 152)

action taken directly by the electricity industry and in this case, EDF:

– to neutralise the regularly rejected effluents, i.e., those coming from the water regeneration, from washing the air heaters (neutralisation of soda), from internal chemical cleanings of the steam generator (neutralisation and decantation before infiltration into the ground);

– to avoid the thermal pollution of rivers, on the one hand by attentively studying with all interested scientific services the consequences of water heating, on the other hand by digging open air canals to allow a natural cooldown (Montereau, Porcheville);

– to avoid accidental leaks of petroleum products (installing of fuel traps wherever leaks are possible).

The legislation concerning noises is even less precise than in all other fields. In spite of the existence of a "technical noise Commission" in the Public Health Ministry, which suggested an acceptable limit of noise in 1961, no general legislation limits noise and EDF has had to fix its own rules. It thus decided:

– to impose certain technical clauses on material constructors whose goods entailled noise emissions higher than 80–85 dB;

– to equip the blowing ventilators, the principal transformers, the safety valves and other noisy installations with insonorising material;

– to take particular measure to protect its personnel.

Hence we see that EDF undertook at very early stages a serious series of actions to fight against the pollution and noise that its activities imposed upon the environment, in spite of insufficiencies which we shall mention further on. In this field, EDF's action often preceded even the public authorities' intervention by filling in the holes of an unadapted legislation and by contributing through research and experiments to define new norms. More positively, EDF took a series of happy steps in the industrial aesthetics field, in developing the surroundings of its power stations, even, wherever the site made it possible, the use of its artificial lakes for tourist activities. Here again we find the behaviour already noted for other aspects of the energy policy. But this situation calls for some more general observations on the conditions for mastering the effects of technical progress in developed societies.

The changes under way and their meaning.

The efforts accomplished by EDF obviously do not imply that France has no pollution problems tied to the development of electric energy. At the most it can be said that these problems have not yet become so acute nor

---

Cf. on all these problems: L. HUDRY and A. GIBAULT, "Prevention of Industrial Nuisances in the Thermal Power Station at Porcheville B," *La Technique Moderne* (Feb. 1971), pp. 53–67.

massive as in other countries. But this situation cannot only be judged on elements from the past for profound changes are under way and we must now discuss their direction and their influence.

First, the variations observed in the growth of the electricity industry are not only conjectural. They reflect the progressive integration of electricity in the competitive part of the energy sector.

The electricity market, after having remained for a long time captive, protected and relatively autonomous, sees the frontiers separating it from the fuel market disappearing, both for the domestic and industrial users. On a long term period the notion of "specific use" is demolished incessantly by technological innovations. This competition between all forms of energy can, however, take on different forms, according to whether the development of vertical integration of transport and distribution networks continues (petroleum products, gas, electricity), or whether one of them overtakes the rest.[19] In this latter case, electricity production could either continue to concentrate into very large units and supply the major part of the final users' consumption (all electric), or fraction itself into small units (industrial or urban) at the terminal points of the fuel and above all gas networks (total power). Numerous factors command these evolutions: the development level of nuclear production, above all if this latter is seconded by growing tensions on hydrocarbons, will probably favour the "all electric"; the progress of the internal combustion engine and of the gas turbine, the growing interest of heat recuperation, the increasing gas reserves in industrialised regions... will all play in favour of "total energy." But it is possible that the determining factors will be tied to the environmental policies adopted in industrialised countries.

In the more specific French case, a second group of changes must be underlined. The more and more competitive aspect of electricity has progressively undermined the concept of a state firm, some aspects of which we mentioned above. Forced to fight its gas and petroleum competitors in diverse fields (domestic house users, industrial users-danger of autoproduction), EDF has had to change its policy in several ways:
– taxation (change in the industrial "green" tariff for large electricity consumers and more flexible commercial relations with clients),
– relations with contractors who are left much more free in the material design and technical progress,
– supply in primary energy more and more commanded by the "least-cost" criterion, etc.

This evolution, by no means over, tends simultaneously to liberate the corporation from restricting administrative ties and to allow it to abandon

19. Cf. General Plan Commissariat, Prospective Group, August 1970, 122 p.

its responsibilities towards other sectors of the economy, in the name of a very broad (even "deformed" according to some) concept of a public service. In such a context one could imagine that in the future EDF might not be willing to pay the costs of quality and protection of the environment if those costs largely exceeded those imposed on its contractors or on electricity auto-producers. The public authorities should thus take back upon themselves the responsibilities which they had up to now left to the state corporation.

Thirdly, this hypothesis leads us to discuss the role of the future public policies. In France the '60s were marked by an acceleration of the observable evolution since the last world war. The sectorial interventions (basic industries, key industries, spearhead industries) principally designed in terms of growth, and whose coherence was only at quantifiable economic relations (investment, employment, foreign currency...) are being progressively replaced by more global actions which only indirectly act on production conditions (science and techniques, professional training, site planning, exterior markets, etc...). The industrial firms have visually seen their power to orientate the growth of the national economy be reinforced, without the public authorities being able to satisfactorily define and even less to apply, policies ensuring the coherence of sectorial growths, not only on the economic activity level but also on the level of the whole society. A good example is given by the problems which are beginning to be discussed about the environment. They imply that, if one wishes to avoid certain dangerous mishaps which the USA is now tackling,[20] long term aims should be rapidly fixed for the electricity industry for everything which concerns its relationships with the environment in the broadest sense (occupation of space, distribution to users, levels of pollution and nuisances, etc.). But it is also clear that such aims could not be defined outside a long term energy policy foreseeing the place of electricity in the total power balance and outside an environment policy covering all pollution and nuisance problems.
The difficulties of such a project are even accentuated for an open economy integrated into the rest of Europe.

We cannot stop here when the problem under discussion is dependent on industrial allocation on a world scale and on its concentration in the developed countries. This widening of the horizon multiplies naturally the number of hypotheses to be explored, but we believe that it would be unsafe to neglect some of them "a priori." This in any case calls for a rapid progress of social sciences in order to master the consequences of technical progress which no-one any longer thinks of judging only in quantitative growth terms.[21]

20. *Le Monde*, "Contested Nuclear Power Stations," 28 April 1971.
21. Thus the necessity to rapidly perfect the methods of technological forecasting, all aspects of systems analysis, the planning and simulation techniques applied to social sciences. Several departments of the IREP-Grenoble are now working in these directions.

Chapter VIII A

ELECTRICAL ENERGY NEEDS AND
ENVIRONMENTAL PROBLEMS IN POLAND

by *Jacek Janczak*

The Substance of Electrification Policy in Poland

Economic development in Poland is directed by the central planning system, which aims to foster harmonious growth of the economy through optimal use of national resources. Practically all industry, railroad transport, and about 15% of cultivated land belong to the state; the remaining 85% of farmland is in private hands. Electrification of the country, one of the key factors in industrial and economic development, is given high priority. Plans for an increase of electrical energy production are based on the balance between increases of industrial, transportation, community, and agricultural power demands and the production capabilities of the energy system. Finances for development of energy needs are provided by the National Economic Plan.

Intensive development of industry is a desired goal of state policy. During the last ten years the average annual increase of industrial production has been 8.3%. There was a corresponding increase of electrical energy production. In the years 1950 to 1960 it was 11% to 13% per year, and from 1960 to 1970 was between 8.3% and 11% annually.

Disposition of new industry centres, town agglomerations, main transport lines, and so on are defined long in advance of the physical development of the country. Assumed development trends in industry and agriculture and an increase of individual consumption connected with a higher living standard allow determination of future electrical energy demands. The expected trend in the growth of the country's energy needs assumes that the use of electric energy will double approximately every ten years.

Projected Electrical Energy Production in Poland

Year	1970	1975	1980	1990	2000
Production TWh	65	96	130–140	260–270	500
Consumption per Inhabitant kWh	2000	2800	3400	6300	11000

Polish energy production policy is based on the nation's own sources of raw materials. Since coal is abundant 98% of the electricity is produced in thermal power stations. In 1968 63% of these were coal-fired and 37% brown-coal fired. Because appropriate sites are unavailable, only 2% of the country's electricity is produced by hydro-energy. Another determining factor is the convenient geological location of very large coal resources and a highly developed mining industry. These conditions make possible an extremely low production cost of coal.

Several methods are being employed in an attempt to obtain better fuel economy. In new urban developments heating is supplied by a district or town system which obtains heat and power directly from large power stations in Warsaw, Lódz, Kraków, Gdánsk, Bielsko, Wroclaw, etc. There has also been a trend toward electrical accumulation heating, where enough load is used at night and other low-load times to keep large units turning, thus avoiding the costs of shutting down. Preferential rates are given consumers using this electricity. These policies, it is hoped, will help eliminate air pollution in towns and industry agglomerations and still provide a large power concentration.

At present, the capacity of the biggest power station in Poland is 2000 MW. Such stations are based on unit systems and have standard 200 MW units designed for less than critical parameters. Planned capacity of new power stations will be 2500–3000 MW. This capacity defines the limiting value for permissible emission of air pollutants.

Low calorific value and residual fuels are used in large power stations, because in this way it is possible to use these fuels efficiently while minimizing their effects on the environment. These stations are equipped with very high smokestacks (200–300 meters) which allow dispersion of air pollutants at great height. Power stations designed for burning low calorific value and residual fuels are situated near mines because otherwise the cost of fuel transport would be higher than that of the transmission of energy. These power stations operate primarily with recirculating water cooling systems. Remaining power stations are situated in energy consumer centres and it has been possible to find enough water in large rivers and lakes to operate open cooling systems.

The present installed power capacity in Poland exceeds 14,000 MW. Assuming the doubling of installed power every ten years, the following increase of total installed power in Poland can be estimated.

Year	1970	1975	1980	1990	2000
Power MW	14 000	20 000	29 000	55 000	100 000

Coal is and will remain the dominant fuel in Poland in the foreseeable future. In 1969, 135 million tons of coal, 31 million tons of lignite, .44 million tons of oil, and 3.92 billion Nm^3 of natural gas were produced. As the amount of oil refined within the country increases, residual oil will be directed in limited amounts to the power stations. It is also expected that newly explored natural gas fields will supply gas for buffer burning, primarily in heating power stations. In comparison with the quantity of coal, however, the amounts of these fuels will be insignificant. To balance the load in the system, the capacity of pumped-storage power stations is being expanded. Taking into account low production cost of coal and its very large resources, the commissioning of the first atomic power station will not take place until about 1980.

The state favours the intensive electrification of agriculture both from a social and economic point of view. The extension of the electric grid in the countryside is carried out with the help of State grants. So far, electric energy consumption in rural areas is small, amounting to about 2.5% of the total. A rise in rural consumption of energy is expected along with the increase of electrical equipment and machinery ownership. At present, more than 85% of all farms are connected to the mains.

Transmission System

All power stations in Poland are connected to the national grid. Transmission lines operate on 400 kV and 220 kV, while the distribution network is on 110 kV and lower voltage. The length of lines in 1969 was:

kV	400	220	110
klm	320	5149	16651

The Polish national system is interconnected and cooperates with Czecho-slavakia, the German Democratic Republic, and the USSR, forming the East-European system which has a total peak load capacity of 60,000 MW. Expenditure on grid extension rises proportionately to the development of the power system.

Electrical Energy Distribution Policy

In order to prevent waste of electrical energy, the State Inspector's Office for Fuel and Energy Economy was established to advise the Minister of Fuel and Power. This Office fixes the limits of energy demand for various industrial plants, depending on their production capabilities and on the type of installed equipment.

To encourage distribution of the demand for electricity, preferential tariffs are set for energy used outside peak hours. A typical example of buffer action is to switch on carbide furnaces at low-load periods, mainly at night, and to switch them off at peak hours. Individual consumers have no restrictions on energy demand, but get a preferential tariff for using accumulation heating.

Impacts and Consequences of Existing Policy for Various Domestic and International Interests

Optimum consideration of all interests important to the country's development is essential to Poland's economic plan. Basing electrical energy production on the country's own fuels frees the national economy from the vagaries of world fuel prices. The creation of the common power system by the socialist countries is advantageous to all partners as it allows more efficient use of power reserves within the system.

The balancing of the supply and demand of electrical energy within the national grid supplies industrial power and makes small and uneconomical condensing stations unnecessary. With the introduction of regional heat and power management, a central technological steam supply was made possible in a number of factories. For example, in Lódz several textile factories are supplied with steam from a large heating plant. These are but a few of the ways the country plans for economical production of electrical energy, fuel economy, and important lowering of air pollution.

The preferred method of heating homes is accumulation heating, which uses off-peak electricity. The cost of direct daytime use of electricity for heating is almost prohibitive. It is higher, in fact, than the cost of directly burning coal or coke. In some suburbs where electrical accumulation heating is widely used, a definite lowering of smoke emissions has been recorded.

The Polish program for the creation of "no smoke" areas includes the introduction of district heating, gas heating, and electrical accumulation heating.

The intensive electrification of rural areas carried out by the State is an important integration factor between town and country populations. It equalizes the living standards of farmers and town-dwellers through the dissemination of contemporary civilization. These processes are similar all over the world, but are more prominent in countries with low population densities.

Construction of large, modern power stations, well equipped with highly efficient electro-precipitators and very high smokestacks, provides better protection to the environment than would smaller and more numerous stations. This can be more clearly seen where fuels of high sulphur content have to be burned. At this time it is impossible to avoid using high sulphur content fuels, because this would mean slowing down the rate of economic development. However, all possible means are being undertaken to diminish the harmful effects on the environment. The expansion of mines producing high sulphur content coal has been stopped, with the aim of limiting its production in the future.

With the development of the national high voltage grid, a new problem arose—that of cutting down forests to allow the construction of new electric lines. New strict regulations have been recently introduced to protect the forests.

The effects of waste water discharged into rivers and lakes can no longer be ignored. Though energy policy is the responsibility of the Ministry of Mining and Power, such specialized government agencies as the State Inspector's Office for Fuel and Energy Economy and the Central Water Resources Administration work together to develop policies which will obtain the desirable economic results while taking into account the best possible ways of protecting the environment. The current policy of constructing new large power stations is an example of this cooperation. These same agencies are responsible for supervising the concluded agreements.

In conclusion it can be stated that within defined geographical and economic conditions, energy policy in Poland is based on realistic premises and pays respect to a great number of important factors in the protection of environment.

Policy Towards Research for New Energy Sources and Technologies

All research in Poland is financed by the State. The national research plan is a part of the National Economic Plan, which reserves finances for research programs. Most research is supported by the "Technology Development

Fund", which, in turn, receives a percentage of the income of national industries.

The determining factor of the energy research policy is the existence of very large coal deposits. On the one hand, the main research task is to guide power station technology towards higher efficiency and lower costs of energy production (large units, higher parameters, special materials, optimization of the power system, better management, and so on). On the other hand, the utilization of coal in power stations makes it necessary to devote attention to research on environmental protection problems (desulphurization of combustion gases and coal, utilization of PFA).

Future erection of nuclear power stations is an important factor in the abatement programme against air pollution. This has inspired a broad research programme in the field of nuclear power. In this area, new technology research is being carried out within the framework of international cooperation.

The completion of such expensive investigations, if carried out separately, would take a very long time to obtain results on an industrial scale. To avoid dispersion of the means for research, efforts are being made to concentrate on the basic problems of the energy policy.

Instruments of the Country's Energy Policies

Within the planned economic system, energy policy is carried out by first determining the tasks to be accomplished; means for their completion are then provided by the State. Trends in energy production and the means for investments are determined in 5-year National Economic Plans by an act of Sejm (Parliament). In accord with the 5-year plan, the Government establishes yearly national economic plans, which are the basis for the opening of bank credit for new investments.

The income from national industry enterprises, as well as taxes, are paid to the Treasury. Prices and rate policies are set primarily to prevent the waste of energy and fuel. The same purpose is served by the limits of maximum permissible loads and energy consumption imposed by the Inspector's Office for Fuel and Energy Economy. All excess demand is penalized by increasing the unit price.

Most research is financed by grants from the State Technology Development Fund. This often includes the costs of pilot installations and prototypes. Development work carried out to improve existing technology is usually financed within the working capital of the enterprises, and this expense is included in production costs. Basic lines of technological development are determined by the National Economic Plan.

In conclusion, one can say that the instruments of energy policy are:

a) State investment credits,
b) Prices and tariffs, and in some cases load and consumption limits, established to prevent the waste of energy and fuel,
c) State grants for important research work,
d) Procurement of fuel supplies. To prevent excessive consumption of imported fuels (fossil fuels), a high price policy is applied.

The Energy Policy-making Process

Formulation of energy policy corresponds to the general trend of the national economy, which is directed by political forces. Basic policy decisions for economic development, energy included, are made by the leading party's (Polish Workers Party) congresses. The national development plan is prepared by the Government Planning Committee in cooperation with the ministries (branches). Political and economic decisions are based upon results of studies and investigations as well as long-range plans prepared by industrial research institutes, the Polish Academy of Sciences, and universities.

The national energy plan is prepared by the Polish Generating Board which is supervised by the Ministry of Mining and Power. The energy development programme consists of a long-range plan for about 20 years, a 5-year plan, and a yearly plan. Both the 5-year and yearly plans are working programmes and are included in the national economic plans. These plans are based on research conducted by the Institute of Energy, Consulting Engineers (ENERGOPROJEKT and ENERGOPOMIAR), and other organizations.

Consulting service is provided by the Scientific and Technical Council of the Generating Board, the National Energy Council, the Minister of Mining and Power, and the Polish Energy Committee, formed within the Polish Technical Association. The main items of the energy plan are:

a) A report on the availability of fuel prepared by the Ministry of Mining and Power,
b) Delivery possibilities of equipment for new power stations and grid extensions provided by the Ministry of Heavy Industry and the Ministry of Machine Production,
c) An evaluation of the necessary construction and fitting work by the Ministry of Construction,
d) Plans and designs for new investments drawn up by consulting engineers (ENERGOPROJEKT).

The present policy for locating new power stations is a product of the national physical development plan, in which the locations of consumer centres are determined. The plan is prepared by the Planning Committee of the Council of Ministers, including the Institute for Physical Studies and Plans in cooperation with regional authorities and committees of the Polish Academy of Science (e.g., Man and Environment Committee, Physical Economy Committee, and so on). Selection of new sites is determined only after consultation with the Central Water Authority which evaluates water demand, water and air pollution, the Ministry of Transport which is concerned with fuel delivery, and local authorities who decide on the exact location of the new station. Approval of all these organizations is required before construction can begin.

Criteria Shaping Electrical Energy Policy

The primary task of State policy is determining the most effective utilization of all means in the national economy to obtain orderly growth and yet acknowledge the importance of environmental protection.
Electrical energy policy has broad economic results as its general criterion. Here the determining factor is the problem of fuel.

It has been mentioned that by basing the energy balance on inexpensive fuel produced within the country, it is possible to avoid surprises connected with fuel price fluctuations on the world market. This consideration and the large export of coal play a significant role in Poland's balance of payments.

An important social policy goal has been full employment of miners, who have a long and honourable tradition in Poland. Indeed, the employment problem has a strong influence on plans for industrialization of various regions.

The growth of industry has produced harmful effects on the environment. Expansion of factories, agglomerations, transport lines, industry installations, and dumps into forest and agricultureal areas is on the increase. The intensive development of industry in the Cracow-Silesia region and the accompanying growth of air pollution has affected the surrounding area and was especially detrimental to the forests. A survey conducted in 1967, indicated that 240,000 ha of woodland, or three percent of the total forest area in Poland, was damaged by air pollution. Because of soil conditions and geographic location, most of Poland's forests are coniferous and thus particularly sensitive to air pollution. In recent years a further increase of damage in forest areas has unfortunately been registered.

At present the problem of environmental protection has a dominant influence on siting policy for new power stations and on the future direction of mining industry development. Many factors are considered when the need for a new power station becomes apparent. Capacity of new power stations is limited by projected air pollutant emissions. This is directly related to the pollution content of the fuel, sulphur compounds being recognized as the worst offenders. The necessity of using low sulphur content fuel on a larger scale caused the mining industry to increase the production of this type of coal. Moreover, research is being conducted in an attempt to develop new methods of desulphurization of combustion gases and coal.

An additional question which must be considered when studying site locations is the low water level in Polish rivers. This limits the possibilities of constructing power stations using open cooling systems in the upper and middle reaches of rivers. This is important in fuel production regions and in some energy consumption areas. A positive factor connected with the discharge of waste heat into the rivers is that it counteracts floods which occur at the time of ice breakups and are caused by ice floats blocking the rivers. This is taken into consideration where rivers are relatively shallow and flooding causes large losses each spring. The power stations, by discharging waste heat, are causing earlier melting of ice and prevent or limit the floods.

Rapidly increasing occupation of agricultural and forest areas by developing industry has caused the Government to issue strict regulations ordering new industrial plants to locate on waste or poor quality land. Protected areas such as national parks, reserves, health resorts, recreation areas, good arable land and forest areas have been protected in the long-term physical plan.

Meeting all demands resulting from consumer needs, system and transport requirements, widely acknowledged environmental pollution, employment policy, and economic development of various regions in the country with the existing fuel resources demands broad site location studies and analyses of alternative solutions. For example, in determining the location in Central Poland of a new power station of 2500–3000 MW final capacity, fourteen site alternatives with thirty-four technical variants were studied. Such studies are carried out by the Central Generating Board in close consultation with the Central Water Resources Administration, Ministry of Forestry, and local authorities. It can be said that environmental protection is now beginning to play as important a role as system considerations in the shaping of electrical energy policy. Some problems relating to environmental protection have become subjects of legislation, in particular the "Water Law," "Clean Air Act," and Government regulations concerning protection of agricultural land, forest areas, and so on.

In Poland, electrical energy policy is only one part of the overall energy policy. Important roles are also played by gas supply policy, heat supply policy for technological purposes (steam supply), and heating of dwellings (hot water supply). The State Inspector's Office for Fuel and Energy Economy recommends, as a rule, construction of large heating power stations, situated in one of the industrial plants, supplying steam and hot water to the entire region. Development of these heat sources takes place with the financial participation of all interested parties. This policy is carried out in cooperation with the air protection authorities, since it allows control of emission of air pollutants in large agglomerations.

According to our experience, environmental protection, emission of pollutants, discharge of hot water, and damage to forests and agricultural areas should not be treated as separate problems. Optimum solution of these problems is possible only when the environment is considered in a comprehensive manner within the full context of the country's economic development. As an example, a heating power station can improve the pollution situation in a whole region because its presence allows the closing of small and low emitors which are far more onerous than a plant well-equipped with electro-precipitators and high smokestacks or an industry utilizing pulverized fuel ash for the production of building materials.

The Capacity for Choice and Change in Energy Policy

The future of energy policy and environmental protection in Poland brings to mind several questions:
a) Will energy policy and the criteria shaping energy policy-making change during the next three decades? New geological discoveries could cause a considerable increase in the resources and production of gas, making possible a more flexible economy based on non-sulphur fuels. Nuclear energy will be introduced about 1980, and by 2000 will represent a considerable position in the production of electrical energy. Coal heating in individual dwellings will be replaced by gas, oil, or electric heating or by heating from the power station system, thus decreasing the emission of air pollutants. After 1975, it is predicted that the demand for low-sulphur petroleum will grow in order to ensure the supply of low-pollutant fuel for the boiler-houses located in agglomerations and for the power stations working in the alarm system. We can presume that in the coming three decades the refinement processes of fuels and combustion gases will be made considerably more efficient. This will enable their utilization in industry and will reduce sulphur compound pollution. It can be expected, however, that nitrogen oxides will still constitute a considerable danger.
b) Do other policy alternatives exist? At present only two approaches to the development of energy seem plausible. Current policy is based on Poland's own raw materials. Future development assumes the partial import of raw materials mainly as an aid in abating pollution. A considerable expansion of nuclear energy facilities is also foreseen. In the light of Polish predictions of the future world demand for coal, the utilization of other fuels within the country will not limit mining industry development.
c) Will a changing technology affect the role of the Government in elec-

trification policy? In Poland all branches of industry, including the production of power, were nationalized after the Second World War. By nationalizing the power industry, the Government has financed and made possible the considerable research, development, and modernization of large power units which provide concentrations of electrical energy capacity.

Chapter VIII B

LONG TERM PLANNING IN POLAND

by *Pawel Jan Nowacki*

Organisation

The supreme body for planning in Poland is the State Planning Commission in which all planning activities are centralized. This commission is attached to the Council of Ministers and is responsible to the Government. The commission—relying on projects conceived by different bodies—prepares a central plan for all investments and for manpower on a five-year basis and detailed plans for every year.

The Government may present the plans to the Polish Parliament (Sejm) which after discussions in commissions votes the plans at its plenary sessions. The ministers of the Government are responsible for the execution of plans in the domains of their respective activities.

Besides the State Planning Commission there is a State Committee for Science and Technology, whose chairman has the rank of minister and is a member of the Council of Ministers. The task of this committee is to advise about allocation of funds for research and development and to supervise spending in all branches of the national economy.

Basic research is carried out chiefly by the Polish Academy of Sciences. There are various committees engaged in long term planning on selected subjects, such as: Man and Environment, Nutrition of the population, The safeguard of nature and its resources, Energy, The peaceful utilisation of atomic energy, etc.

As far as electrical energy is concerned, the main Government Authority is the Ministry of Mining and Power, which incorporates the Central Energy Board, responsible for the production and distribution of electrical energy in Poland.

The methods used in long term planning are:

extrapolation methods especially for short term planning;
correlation methods based on the Gross National Product (GNP) indexes.

All plans are aiming at a maximum utilisation of national fuel resources.

As to liquid and nuclear fuels, imports have to be included. The imports require consultations within the framework of the Council for Mutual Economic Assistance (CMEA) in which Poland is a partner.

Lines of development

Poland disposes of various resources of energy. In order of importance may be mentioned:

– hard coal, about 85 GT to a depth of 1000 m, irrespective of new discoveries in the Eastern agricultural district of Lublin;
– brown coal (lignite), about 40 GT to a depth of 400 m;
– natural gas;
– crude oil in rather limited quantities;
– hydroelectrical power, exploitable to 6 or 8 TWh/a.

Coal resources are concentrated in Upper and Lower Silesia, whereas brown coal is mined in the centre of Poland. It mainly serves the production of electrical power.

An increasing demand for natural gas is anticipated for urban heating as well as for electrical power generation.

Table 1 represents consumption and production data of 1965, with estimates for 1975, 1985 and partly for 1995.
The consumption of electrical energy in Poland is expanding at a steady rate. The increase in the next five year interval is expected to be about 45%. The growth so far does not show a tendency of slackening; as a consequence appropriate investments are needed in plants and in transmission and distribution lines.

Table 2 gives yearly figures, indicating the population growth in Poland and the development of electrical energy consumption and production from 1950 to 1970, together with strict extrapolations to 2000. Graphs are shown in figure 2 to a semilogarithmic scale.

Figure 1 gives trends of the installed capacity of production of electrical power in Poland and in the world as a whole, from 1965 to 1985. According to present calculations this total capacity increases at a growth rate slightly above the 7,07% per annum which leads to a doubling time of ten years. In evaluating the prognosis one should keep in mind that the installation of 2500 GW power capacity—the assumed increase of world capacity from 1970 to 1985—requires an investment of $ 250.10^9—($100 per KW)!

In figure 2, also, the growth of nuclear power production is shown as computed by the International Atomic Energy Agency (I.A.E.A.). Only plants of a proven technology, i.e. light water reactors, heavy water reactors and high-temperature reactors have been taken into account. Fast breeders are not included. Their importance, however, will be more and more apparent after 1985.

In Poland a diminishing yearly growth rate of electrical energy consumption may be expected, from 9% during the years 1966 — 1970 to 7,9% in 1971 — 1975 and to about 7,5% in 1976 — 1980. This tendency results from a certain saturation of heavy industries or, in general, of the great consumers of electrical energy. Also, there is an increasing influence of direct energy supply from sources like natural gas and liquid fuels and in addition a widening use of thermal energy from district heating plants.

The share of the industry in the usage of electrical energy is anticipated to fall from 75% in 1970 to 73% in 1980. A higher percentage will be needed for domestic and agricultural purposes.

To render certain that production capacity will be at the required level, new power plants of a capacity of 16000 MW have to be built from 1971 to 1980 of which 1800 MW capacity would come from hydroelectric plants.

It is expected that nuclear power plants will play a certain role after 1980, the first plant being of a light water type (PWR) at a capacity of 1000 MW. Estimates are that in the year 2000 a total nuclear capacity of 20000 MW and an overall capacity of about 114000 MW will be needed.

Fig. 1. Installed Power Capacity (GWe). 1965÷1985.

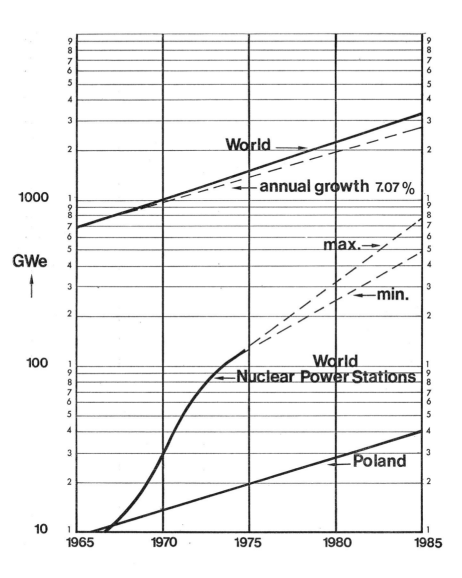

170

Fig. 2. Energy Prognosis for Poland.

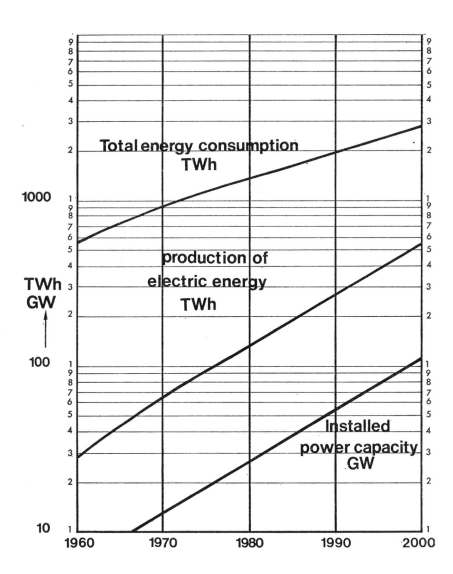

Table 1. Energy Production and Consumption in Poland between 1965 and 1995.

Nr.	Description	Unit	1965	1975	1985	1995	Remarks
1	Population	10^6	31,6	35,5	39,6		a)
2	Consumption	Tcal	653	1000	1400		a)
3	Consumption	TWh	757	1160	1620		a)
4	Total production	TWh	880	1260	1405		b)
5	Coal	TWh	788	1070	1150		b)
6	Lignite	TWh	52	89	89		b)
7	Crude oil	TWh	4,06	19,7	65		b)
8	Natural gas	TWh	13,45	62,7	97,4		b)
9	Peat and wood	TWh	22,7	11,6	9,3		b)
10	Hydroelectric prod.	TWh	0,93	2,3	8,1		b)
11	Import	TWh	77	145	288		b)
12	4 + 11	TWh	957	1405	1693		b)
13	Export	TWh	198	209	62,6		b)
14	3 + 13	TWh	955	1369	2246		a) b)
15	Saldo (4 — 3)	TWh	+ 123	+ 100	— 215		c)
16	Electrical energy consumption netto	TWh	36,3	82	164	326	a)
17	Electrical energy production gross	TWh	43,8	96	193	384	a)
18	Installed power capacity	MW	9670	20300	40800	81000	c)
19	Utilisation time of installed capacity	h/a	4530	4720	4730	4730	c)

a) A. Kopystianski et al.: WPC VII, doc. AL90.
b) C. Mejro: Podstawy Gospodarki Energetycznej WNT Warsaw 1968.
c) P. J. Nowacki: The Development of Nuclear Power Stations in Poland, Warsaw 1968.

Table 2. Consumption and Production of electrical energy in Poland

Nr.	Description	1950	1955	1960	1965	1970	1975	1980	1985	1990	1995	2000
1	Consumption gross (TWh)	8,82	16,47	27,27	40,07	61,1	89,6	127,0	178,2	250	354	500
2	Consumption netto (TWh)	8,03	14,86	24,43	36,3	55,8	82,2	116,8	164,2	231	326	460
3	Production gross (TWh)	9,42	17,75	29,30	43,80	65,3	96,0	138	193	272	384	542
4	Installed Power Capacity (GW)	2,74	4,18	6,32	9,67	14,13	20,3	29,2	40,8	57,5	81,0	114
5	Population in Millions	25,03	27,55	29,8	31,6	33,4	35,5	37,6	39,6	41,7	44,0	46,5
6	Consumption per capita and year kWh/a. cap.	324	544	827	1145	1680	2310	3120	4140	5540	7420	9900

Part III

FUTURE STRATEGIES AND APPROACHES

Chapter IX

AN AMERICAN NATIONAL ENERGY POLICY?

by *Robert M. Lawrence*

The United States has neither a general national energy policy nor a specific electrical energy policy. A number of reasons account for this state of affairs. In the first place there is the political system which features a division of political authority not conducive to national planning between the central government and the governments of the 50 states. Further, within the central government, and all the state governments, political authority is dispersed between executive, legislative and judicial entities. And, within the executive branches, generally the largest government entity at any level, there are found a host of departments, agencies, bureaus, "independent" commissions, etc., many of which have strong ties with various and opposing pressure groups in society.

The First Amendment to the Constitution, which guarantees "free speech", "free press", "free assembly" and "free petition of the government" enables any group or person who eschews violence the opportunity to comment or act upon political and economic matters. Although as Robert Dahl has mentioned many Americans fail to use all the political power which is legally theirs.

Then there is the traditional capitalistic view in America that energy, like most goods and services, can be best provided and marketed by units in the private sector of the economy which respond as they see fit to the demands of the market. Thus the coal industry may pursue certain activities while the petroleum companies may follow others. In association with the capitalistic tradition is the general distrust, even fear, of government planning and centralization of decision-making which appears to be Socialistic, or Communistic. Although lessening now, the intense U.S. preoccupation with anti-Communist policies during the last quarter century still fuels much hostility toward any government activity which hints at collectivism which, it is said, would rob Americans of their political and economic freedoms.

Lastly the lack of a national energy policy probably reflects the preference of the most powerful energy pressure group, one of the most influencial of any American pressure group, the petroleum companies. The point here is that the petroleum companies, or as some are calling themselves the energy companies, can probably better enhance the self-interest aspect of their operations when government policies concerning them are weak or non-existant, than when the government operates effectively to organize energy production for the benefit of all citizens. Regarding the petroleum companies, one might argue that instead of there existing a national energy policy, there has developed some bits and pieces of government activity which assists the petroleum companies more than the general population. A case in point would be state and federal government regulations which apparently serve the purpose of limiting foreign competition and supporting the stability of the market. All of this means that if the U.S. does develop a national energy policy, the very interests, private and public, with a stake in such matters must be generally won over to the emerging government activity. Or, as some social scientists would put it—support for a truly national energy policy must be won by development of a positive non-zero sum game arrangement.

However, the process has begun which could lead to the development and implementation of such energy policies in the next few years as bits and pieces of existing governmental activity, together with newer concepts, are evaluated relative the kinds of futures from which America may choose.

Three official points of departure for the development of a national energy policy are: 1) President Nixon's special message to the Congress of June 4, 1971 on energy; 2) the President's proposed reorganization of the Cabinet resulting in the placement of all energy-related Federal entities in a new Department of Natural Resources; and 3) the hearings on a national energy policy begun in the summer of 1971 by Senator Henry Jackson, Democrat from Washington.

The interface which could develop between the U.S. efforts to develop a national energy policy and the activities of other nations may be suggested by examining the frame of reference against which the U.S. will argue out its energy objectives and the alternative strategies for obtaining them.

Currently the emerging debate about a national energy policy involves two major and to some extent opposed considerations. On the one hand a majority of Americans seemingly believe that the substained rise in the U.S. standard of living ought to be maintained, and that this goal includes a greater use of electrical energy. On the other hand, there is concern about environmental degradation caused in part by the various energy production methods now in use.

One may easily recognize these two considerations in the current literature, both popular and scholarly. One may also readily perceive the tension which has developed between those who can be classed under the "more is better" heading vis-a-vis those who now prefer certain environmental values over other material ones. Now it appears possible that the two differing considerations may be brought into a more harmonious relationship by development of alternative power sources which will produce an impressive amount of electricity which is both *relatively* cheap and *relatively* harmless to the environment.

With the above thoughts in mind it comes as no surprise that a major component of what could become a U.S. national energy policy is the consideration being given by the Nixon administration to making the Liquid Metal Fast Breeder Reactor a national technological goal. Part of the overall L.M.F.B.R. decision process is the effort by those who would convince the government that fusion and solar energy technologies ought also to be incorporated in a national drive to develop guaranteed sources of electrical energy acceptable to most Americans.

While the debate grows and spreads over how to strike the best balance between the demand for more electrical energy and greater environmental protection, other considerations are developing below the perception of the general American public. These, which are being noticed by various government and scholarly groups could well be even harder to solve than the more power/better environment matter. As suggested below the inherent conflict between these considerations could be substantial.

These considerations include:

1. The idea that scientific/technical achievements of great import to all, which have been largely developed through expenditure of public funds and the efforts of government employees/officials, should belong in some fashion to the public. An associated idea is the one which holds that scientific/technical projects of great value to all human beings, which require money and talent in quantities only possessed by a very few if any nations, ought to be developed through international cooperation.

2. A belief that capitalism "made this country great" and that therefore any interference with the system is both economically foolish and politically dangerous.

3. The fact that impressive civilian-oriented science and technology often are perceived as having significant military implications.[1]

4. A continued decline of the U.S. economic situation, particularly in regard to the balance of payments, which should stimulate the search for ways to strengthen the U.S. economic position.

One can perceive the opportunities for argument which the above considerations may generate within the U.S., and some possible ramifications for international affairs which could result from the U.S. debate and decision.

1. C. F., Samuel T. Cohen, "Tactical Nuclear Weapons and U.S. Military Strategy," *Orbis* (Spring, 1971); and C. L. Sulzberger, "Solving an Ugly Dilemma," *New York Times*, November 15, 1970.

Let us now turn our attention to three "New energy" possibilities for alternative electric energy production, Liquid Metal Fast Breeder Reactor, Fusion Reactor, Solar Collecting Facility. The three appear particularly interesting in terms of the more energy/better environment problem which is perceived by many to be the central issue in the creation of a national energy policy. The three also could be significant relative to the less clearly perceived problems of public versus private power, weapons implications, and trade competition.

The Liquid Metal Fast Breeder Reactor as an Alternative Source of Electric Power

The technology seems at hand which would make possible development of a pilot LMFBR and then commercial use of these reactors on a wide scale. In fact President Nixon has pledged government support for completion of a demonstration reactor in the range of 300 to 500 megawatts by 1980. Thus, unless unexpected technical problems develop or political feelings adverse to breeder reactors escalate considerably, it appears that the most immediate American step toward resolving the more power/better environment problem with "new energy" possibilities will take the form of LMFBR construction.

Regarding the dangers to the environment from thermal and nuclear radiation connected with LMFBR's, the government has made some rather firm statements. For example, the Atomic Energy Agency is on record as saying the fast breeder, "provides the most efficient means of producing energy with minimum effects on the environment."[2] For his part President Nixon recently stated, "We have very high hopes that the breeder reactor will soon become a key element in the national fight against air and water pollution. In order further to inform the interested agencies and the public about the opportunities in this area, I have requested the early preparation and review by all appropriate agencies of a draft environmental impact statement for the breeder demonstration plant in accordance with Section102 of the National Environmental Policy Act. This procedure will ensure compliance with all environmental quality standards before plant construction begins."[3]

Since the decision has virtually been made to press on with the breeder development, essentially on the grounds that breeders will satisfactorily meet the more power/better environment requirements, only those con-

2. U.S., Congress, Joint Committee on Atomic Energy, *Hearings, A.E.C. Authorizing Legislation*, Fiscal Year 1971, 91st Cong., 2d Sess., 1970, p. 1205.

3. Richard M. Nixon, *The President's Energy Message, June 4, 1971* (Washington: Government Printing Office, 1971), p. 4, also p. 226 hereafter.

siderations not immediately perceived by the public remain to be threshed out in the political arena. The matter of whether, or in what fashion, the breeder reactors primarily produced by government personnel expending tax funds should be owned or otherwise controlled by the American people has hardly been raised, much less settled. There are however some interesting facts upon which to launch a preliminary discussion.

For example, there is the testimony before a congressional committee of a U.S. Atomic Energy official which suggests that the traditional expectations made of captalism, (private money and initiative will assume risks and follow imaginative paths toward new and useful productive activities) do not in fact occur in regard to reactor research-development-test-evaluation. Further it should be noted that the government is contributing approximately 75 percent of the funding for reactor development (all programs) while industry accounts for about 25 percent.[4]

The question then arises—should the taxpayers pay the bills for development of "new energy" power sources which then enter the economy as units owned by private entrepreneurs. At this point it is suggested that the U.S. could profit by careful study of how energy is developed, produced, and sold by other societies, including Communist nations. It is entirely possible, although not necessarily so, that we Americans could learn something.

On the matter of weapons applications it is clear that an expansion in the use of fast breeders, in the U.S. and elsewhere, will create additional problems of control for the Nuclear Non Proliferation Treaty, and SALT number II, should it develop. Stocks of fissionable materials will be dramatically increased by the widespread adoption of fast breeder reactors, whether they be uranium or thorium users, which means that opportunities and incentives for diversion of fissionable materials to weapons development will be greatly enhanced.

Regarding the U.S. balance of payments, U.S. success in developing a fast breeder reactor could help strengthen the overall American trading position, and assist various nations meet their energy requirements. While such a development could be mutually advantageous to all, one can conceive of situations where U.S. interests might be served to the disadvantage of other nations. In any event there are important implications for international cooperation and competition.

4. U.S. Congress, *op. cit.*, p. 1135.

Fusion Energy as an Alternative Source of Electric Power.

Work upon fusion has progressed to the point, particularly in the United States, the Soviet Union, and in Europe, where responsible scientists are willing to say for the record that fusion will be an important source of energy within two or three decades. Assuming that many scientists from different nations are probably correct in their judgment about the technical success of fusion for electrical production the question remains—what of the environmental impact of such technology? The matter may be examined under the following categories: Air Pollution, Radiation Pollution, Thermal Pollution, Land Use and Esthetic Pollution, Pollution and the Fusion Torch.

A major advantage claimed for fusion powerplants is that they do not require the burning of hydrocarbons, thus avoiding the release of undesirable combustion products into the atmosphere. Fission powerplants also possess this advantage over fossil fueled plants, as do hydroelectric, geothermal, tidal, and solar facilities.

Another important advantage possessed by fusion powerplants, as against fission facilities, is that whatever chance exists for a "runaway" accident regarding the latter does not exist for the former. The term "runaway" refers to the possibility that a fission powerplant which must contain a "critical mass" of fissionable material, could undergo an accident which would cause the fissioning process to proceed at a faster speed and with greater heat accumulation than normal. The resulting fires and explosions (not a nuclear explosion) could introduce lethal amounts of radiation into the local environment. Their spread beyond the immediate environs would be determined by many factors, including the weather conditions.

Recognition of the possibility, however remote, of a "runaway" accident in regard to fission powerplants was indicated in 1957 when the AEC issued a report setting the best and worst case parameters for fission reactors. Depending upon the circumstances, "...the theoretical estimates indicate that personal damage might range from a lower limit of none injured or killed to an upper limit, in the worst case, of about 3,400 killed and about 45,000 injured.

Theoretical property damages ranged from a lower limit of about one-half million dollars to an upper limit in the worst case of about $ 7 billion. This latter figure is largely due to assumed contamination of land with fission products.

Under adverse combinations of conditions considered, it was estimated people could be killed at distances up to 15 miles and injured at distances of about 45 miles. Land contamination could extend for greater distances.

In the large majority of theoretical reactor accidents considered, the total assumed losses would not exceed a few hundred million dollars."[5]

Further recognition of the possibility of a fission powerplant accident is found in the Price-Anderson Indemnity Act., PL 89-210, which provides up to $ 500 million in government indemnity in the case of an "extraordinary nuclear occurrence." The purpose of the act is to reduce the deterrent to industrial utilization of fission reactors associated with the threat of high liability claims should a serious accident occur.

According to two AEC scientists, fusion reactors are inherently incapable of a "runaway" accident.

There is no "critical mass" required for fusion. In fact, the fusioning plasma is so tenuous (even in the "high density" machines) that there is never enough fuel present at any one time to support a nuclear excursion.

Additionally it should be pointed out that a power loss affecting the strength of the "magnetic bottle," or the structural failure in the containment vessel, and/or outer shielding wall would lead to an immediate drop in temperature and density of the plasma, which would stop the fusion reaction. Or as the engineers would put it, the plant would "fail safe." However, it should be noted there is a possibility that the structural failure of the containment vessel due to neutron bombardment, could result in some release of radioactive tritium.

In the event of an external disaster fusion reactors are safer than fission powerplants. The reason is the large amount of radioactive material, fuel and wastes, which a bomb blast, earthquake, or airplane crash could cause to be released into the environment in the case of a fission reactor. In contrast to a fusion reactor, the materials which could be released from a fission powerplant are among the most persistently radioactive of any known.

If tritium is used in combination with deuterium as the fuel for fusion powerplants, there will be by all scientifically based estimates some "leakage" into the environment. How low that leakage can be kept, and what its immediate and cumulative effect upon living matter will be is not yet fully established. Gough and Eastlund state:

"Carefull design to prevent the leakage of tritium fuel from a deuterium-tritium reactor is mandatory. Engineering studies that take into account economic considerations indicate that the leakage rate can be reduced to .0001 percent per day. The conclusion is that even for an all-deuterium-tritium fusion economy the genetic dose rate from worldwide tritium distribution would be negligible."

"A technology-dependent but possibly serious limitation on deuterium-tritium fusion plants could be the release of tritium to the local environment.

5. U.S., Congress, Joint Economic Committee, *The Economy Energy and the Environment*, 91st Cong., 2d. Sess., 1971, p. 112.

The level would be quite low but the long-term consequences from tritium emission to the environment in the vicinity of a deuterium-tritium reactor needs to be explored."[6]

There is a considerable difference between fission and fusion powerplants relative to the radioactive products and end wastes associated with each. In the fission process the fuel, generally isotopes of uranium or plutonium, cannot be fully consumed. At given points in the operation of a fission plant the spent fuel, associated fission wastes, and newly created plutonium must be removed and replaced. This material is extremely radioactive and such materials generally have a long half-life. After removal from the plant the spent fuel and associated residue is shipped to a fuel reprocessing plant (the only operating plant is near Buffalo, but another is being built near Chicago, and a third is planned for South Carolina). At the reprocessing plant the spent fuel and residue are dissolved in acid to facilitate recovery of "unburned" portions of the original fuel and the new fissionable material created by neutron bombardment. The liquid residue from the reprocessing is still very radioactive and remains so for thousands of years. At present the AEC proposes that such residue be converted into non-soluble solids and buried in earthquake free areas. One proposed burial area is a group of abanboned salt mines near Lyons, Kansas, but even this choice of location is not without controversy. During the removal and transportation of spent fuel and associated fission products, and burying of wastes, accidents can happen, safety standards can be too weak, or compliance poor.

The AEC estimates that by 1980 the liquid high level wastes accumulated from fission plants will be 3.5 million gallons, and that this could be reduced to a solid material which would fill a cube 32 feet on the side. The AEC sees no difficulty in disposing of such wastes.[7]

Whatever the dangers associated with end products and wastes from fission powerplants, it appears that similar dangers of fusion plants are considerably less serious. Fourteen months ago the AEC made the following statement, "In a fusion powerplant the final reaction products are nonradioactive. Thus, there would be no radioactive fission product wasted and no pollution released to the atmosphere."[8] Nevertheless, one can argue and cite other AEC literature in doing so, that a kind of end product and waste radiation problem does exist for fusion powerplants.

In the first place the walls of the containment vessel will be subjected to

6. William C. Gough and Bernard J. Eastlund, "The Prospects for Fusion Power," *Scientific American* (February, 1971), p. 63.

7. U.S., Congress, Joint Committee on Atomic Energy, *Selected Materials on Environmental Effects of Producing Electric Power*, 91st Cong., 1st Sess., 1969, p. 45.

8. U.S. Atomic Energy Commission, *Fundamental Nuclear Research* (January 1970), p. 177.

intense neutron bombardment which will, over time, induce substantial radiation in the wall material. It is not yet known exactly what the effect of such neutron bombardment will be on the wall material. However, it is assumed that the walls will have to be replaced at specified periods in order to preclude structural failure. This means that there will be, if not radioactive wastes, what may be called "radioactive junk" resulting from fusion powerplant operation.

Secondly, it appears possible to design what might be called a "fusion breeder" reactor, the end product of which would be radioactive, but not a waste requiring disposal. Such a facility would utilize a portion of the neutrons produced by the fusion reaction to bombard lithium in order to produce tritium, which can then be used to fuel the reactor which produced it. Instead of having to dispose of the tritium, it would be cycled back into the nearby reactor as fuel. Presumably the amount of tritium so produced would be only enough to fuel the reactor, although in some cases more tritium would be produced for shipment to new reactors requiring an initial fueling. In any event there would often be more tritium on the site than was being consumed in the fusion reaction. This situation would complicate the problem of tritium leakage, and contamination in case of sabotage, earthquake, or other external disturbance.

As had been demonstrated, whether the topic is engineering accidents, operational radiation, sabotage and external disasters, or end product and waste radiation—tritium is always the environmental culprit. Herein then is to be found a great part of the confrontation, assuming fusion reactors are technically possible and economically feasible, between the more-power advocates and the groups under the environmental banner. It is here that some of the major political battles will be fought and the decisions made regarding trade-offs between the hazards of tritium radiation and the advantages of fusion power. Only if a means is found to fuel a fusion reactor without using tritium will this problem be completely removed.

Currently operating fossil fuel or fission powerplants produce more heat than can be utilized. Generally the most economical method to dissipate such heat is by using quantities of water. This accounts for the fact that large fission plants are often located near bodies of water.

Thermal pollution of bodies of water used for the cooling of fission reactors has recently become a "hot" political and legal issue, and this trend may continue as larger fission plants are built. After a stiff political fight, centered on the issue of thermal pollution, the New York State Electric and Gas Company decided against original plans to build a fission reactor near Ithaca, N.Y. on the shores of Cayuga Lake. As one might expect, faculty from Cornell University played a prominent role in the effort to oppose the power company. Mrs. Dorothy Nelkin of Cornell has recently prepared a case study of the controversy.

A case is now before the AEC, which centers on the thermal pollution question in regard to a proposed fission reactor at Palisades, Michigan. An adverse decision for the power company would create serious roadblocks for future fission plant construction.

It is not clear at this time whether all fusion powerplants will require extensive cooling arrangements. Those which produce a heat cycle for electrical energy production or other purposes may, because of their size create substantial thermal pollution. Proposed sizes run between 10,000 and 15,000 megawatts, as compared with the 1,000 MWe fission plants now under construction. Of course if the tritium problem is resolved, or society decides to live within close proximity of whatever tritium leakage occurs, much waste heat could be used for such purposes as city heating.

On the other hand it seems to many to be theoretically possible to build a fusion plant which would avoid the heat transfer cycle by direct conversion of the kinetic energy in fusion reaction to electrical energy.[9] The perfection of this technique would remove a substantial environmental argument against the fusion reactor.

The radiation pollution arising from the use of tritium in fusion power-plants, and the possible thermal pollution caused by such plants may be resolved in a fashion which is tolerable to the majority of urbanites desiring additional electrical energy and the other advantages claimed for fusion reactors, but in a fashion which will create a different kind of environmental problem. For example, if environmental concerns prevent fusion plants from being located near urban centers, or upon certain bodies of water, the obvious answer is to site the fusion plant where political opposition will be low or nonexistent.

The present situation of fossil fuel plants operating and planned on the Colorado Plateau offers a kind of analogue based upon the old saw, "out of sight out of mind." A combine, now numbering 24 utility companies, recently built two large coal-fired generating plants, one at Four Corners, New Mexico and one at Mohave, Nevada. The "politics of environment" aspects of this development are fascinating, but one in particular is germane. That is the fact the generating plants operating and planned for the Colorado Plateau could not operate in the Los Angeles basin for they would be illegal due to concern over air pollution, and the resulting regulations. Although conservationists are disturbed, the Colorado Plateau siting is legal, as is the construction of long transmission lines to carry the electricity to the urban areas in California and Arizona, and possibly elsewhere.

9. William Gough, *Why Fusion?*, U.S. Atomic Energy Commission, Research Division (Washington: Government Printing Office, June 1970), p. 297.

There seem at this time to be no particular reasons why fusion reactors cannot be located in areas similar to the barren sparsely populated Colorado Plateau should they be rejected near cities. But, as Barry Commoner reputedly said, "in nature there are no free lunches." The costs of distant siting of fusion plants are of two types. Obviously such siting simply transfers the tritium leakage problem, and the thermal pollution problem from one area to another presumably less valuable in environmental terms. Alternatively there is the environmental cost of building long transmission lines.

Part of the long-lines cost is esthetic, particularly when the lines and supporting towers cross scenic or wilderness areas, or enter metropolitan regions in residential sections. Part of the environmental cost occurs in terms of land made unusable for other purposes because of the presence of high voltage lines overhead. Transmission lines can be buried but the economic costs are high. However, in one instance the economic cost may have to be borne for underground lines with esthetic and land use values accruing as a bonus. The situation could arise with the construction of very large fusion power plants. Today if a generating plant in a regional grid fails, another nearby plant can provide covering power using overhead lines. However in the case of a 10,000–15,000 MWe plant there may not be any need for a nearby plant because the large plant's capacity is sufficient for a large area. Thus in the event of plant failure, power would have to be brought in from very long distances, too long perhaps for economical use of overhead lines where there is loss of energy related to distance travelled. A possible answer, given technical progress would be very long transmission of electrical power in cryogenic cables (buried lines super-cooled by liquid helium in a surrounding pipe). In such a super-cooled condition there is virtually no power loss regardless of distance travelled. But this is expensive, and still in the early development stage.

There is an interesting suggestion, which like the fusion reactor itself, does not yet exist in terms of hardware, that the ultra high temperature plasma of a fusion reaction be used to" ...convert any material to its basic elements or to produce large amounts of ultra-violet radiation." The "Fusion Torch", a term given to the potential application by two AEC scientists (Drs. Gough and Eastlund), could in theory:

> vaporize and ionize any solid material—converting initially complex chemical compositions into an ionized gas consisting only of elements. The elements could then be separated by a variety of techniques. Toxic chemicals could be reduced to their basic constitutents, ores reduced and alloys separated.[10]

10. Glenn T. Seaborg (Chairmain of the U.S. Atomic Energy Commission), "Fission and Fusion Developments and Prospects," Remarks made at Berkeley, California, November 20, 1969, p. 18.

Dr. Farrington Daniels of the University of Wisconsin has suggested a number of uses for ultra-violet radiation produced by a "fusion torch." Among these are large-scale desalting of sea water, sterilization of sewage and drinking water, and possibly the direct synthesis of carbohydrates from carbon dioxide and water.

It should be noted that many of the projected uses for the "fusion torch" are responses to increasing problems associated with pollution in cities. Hence there may be a trade-off problem regarding "fusion torch" usage between siting in or near cities to facilitate application to waste disposal, materials recycling, water purification, sewage treatment, etc., and distant siting (due to tritium contamination or thermal pollution concern) which would then require heavy transportation costs associated with moving garbage, junk, water and so on to the fusion reactor.

There are also some substantial indications that fusion technology will generate important military applications. Of these the foremost currently appears to be what is termed fission-free fusion weapons which are also called the neutron-bomb or truely clean H-bomb.

The fusion stage of thermonuclear (H-bomb) weapons must now be triggered by an atomic bomb. This means long lasting fission products. The third stage of a thermonuclear weapon also produces a quantity of extremely radioactive and long-lived fission products. In contrast a fission-free fusion weapon would not involve the long lasting fission radioactivity of weapons currently in the U.S. stockpile.

While the fission-free fusion idea is not new in the nuclear era, interest in it has increased recently for several reasons. First, it may now be more technically feasible due to "spin-offs" from the peaceful fusion research. Second, from the military point of view fission-free fusion weapons are increasingly attractive because they are "usable" (assuming little or no genetic damage to survivors) by virtue of very low radiation levels and minimized blast effects. In contrast to other nuclear weapons, a fission-free fusion weapon may release up to 80 percent of its energy as prompt radiation. Thus fission-free fusion weapons might be detonated at certain heights above the ground so as to produce a lethal radiation dose to personnel within a third mile, while creating very little residual radiation, and practically no damage to structures. For these reasons those opposed to the fission-free fusion weapons concept sometimes refer to the weapon as a "capitalistic bomb" because it destroys people and not property. Third, the emphasis placed by the Nixon Doctrine upon reduced U.S. manpower overseas suggests fission-free weapons could increasingly be viewed as a kind of surrogate for U.S. troops.

In view of the growing political and economic pressures for a reduction of U.S. forces, a likely place for such deployment of fission-free weapons as a replacement for U.S. troops would be Europe. One reason why U.S. forces cannot be replaced by tactical nuclear weapons now in the inventory is that their use would create too much blast and long-lasting radiation damage.

In opposition to the fission-free fusion weapon is the argument that since such weapons involve relatively cheap deuterium and tritium, compared to the expensive U-235 and Pu-239 required for current nuclear weapons, they would in fact constitute a "poor man's bomb," thus facilitating the proliferation of nuclear weapons. Then there is the general argument, currently rather widely held, that new avenues along which arms races may be run ought not be opened up for a host of reasons found in the arms control/ disarmament literature.

In spite of arguments against development of the fission-free fusion weapon, a problem the U.S. will increasingly have to face, is increased pressure to build such weapons stemming from the fact peaceful fusion research could place them within relatively easy reach.

An indication of the relationship between peaceful research and weapons research may be found in a recent announcement by the AEC that a commercial firm (KMS Industries, Inc. of Ann Arbor, Michigan) had been granted AEC approval to conduct at its expense research upon fusion reactors:

> ...which would involve the irradiation of pellets of thermonuclear material by a high-power, shortpulsed laser. The pellets would be heated to thermonuclear temperatures (millions of degrees) very rapidly and the thermonuclear energy possibly released on a controlled basis in a reactor.[11]

Development of the laser ignition technique could lead to a means to "trigger" the deuterium/tritium for a fission-free fusion weapon without the necessity of using an atomic bomb or using accelerated particle injection or temperature increase by compression. The AEC announcement also suggests more fully the relationship between peaceful and military fusion research:

"...the KMS proposal also indicates certain work would be similar to that conducted as part of classified AEC programs involving principles and techniques associated with nuclear explosive devices. After a careful study of the implications of the proposal, the Commission has concluded that the KMS work can be conducted under classification and other controls... If, in the future, the AEC determined that the work was leading to the fabrication of a nuclear explosive device or a component of an explosive, the KMS program would be ended or redirected toward peaceful applications

---

11. *Fusion Fore Front* (February, 1971), p. 3. This newsletter is published by the AEC of the U.S.A.

at the discretion of the Commission.[12]

Here then the scientist, philosopher, student of government, interested citizen, and government official are confronted with a set of difficult choices, about which conflicting claims are made, which involves much more than what is generally meant by the phrase "environmental implications of technology." For example, should man cancel fusion experimentation directed toward achieving the peaceful advantages accruing to those operating fusion reactors because of the possibility that such work will lead to a new family of weapons? Alternately, can man arrange to develop peaceful fusion technology while avoiding the temptation to cross over into the military applications? Complicating matters is the controversial suggestion that currently stocked nuclear weapons, particularly those in the multi-megaton H-bomb, range," ...may yet induce reason and restraint, if not wisdom, among the nations of the world"[13] because of their catastrophic environmental effects. Assuming for the moment the correctness of the above statement, it can then be argued that development of fission-free fusion weapons, whose use might not be so restrained by environmental and other concerns, could lead to their widespread deployment, employment, even legitimization, hence the current "balance of terror" would be circumvented to some extent. But then what of the view that man is a belligerent creature who is going to fight anyway and therefore weapons ought to be developed which are the most sparing of survivors (assuming little or no generic damage) and the environment. These might be fission-free fusion weapons relative to the currently stocked nuclear weapons, napalm, and high explosive bombs and artillery shells.

The development of fusion reactors and their marketing on the world scene could prove to be an important boon to the U.S. balance of payments situation. Since considerable time will pass before such a reactor is commercially available there appears time for study of the best, in terms of international impacts, method for the U.S. and/or other nations to plan for providing such reactors beyond their borders.

The pattern for funding fusion is much like the early days of the LMFBR program in that the great bulk of the money spent between 1950 and 1971 (approximately $ 350 million) has been government funds. However, with greater hopes now being expressed about the success of commercial fusion, industry is beginning to contribute financially to the research. For example, the New Jersey Public Service Company is contributing $ 50,000. Others are expected to follow. But the question remains—to whom should fusion reactors belong?

12. *Fusion Forefront, op. cit.,* p. 4.
13. Gaylon Caldwell and Robert M. Lawrence, *American Government Today* (New York: Norton, 1969), p. 513.

Solar Energy as an Alternative Source of Electric Power

It has been suggested by A. B. Meinels, the Director of the University of Arizona's Optical Science Center and his wife that the basic technology has been verified which will permit the thermal conversion of solar energy for use in producing huge amounts of electricity. In fact the Meinels suggest that ultimately it will be technically possible, and economically feasible, to construct a "National Solar Power Facility" along the Colorado River in the southwestern United States with a total capacity of 1,000,000 megawatts of electricity. Additionally, they suggested that the waste heat from such a facility could be used to desalinate 50 billion gallons of water per day. The success of the project appears to hinge upon the further development of highly selective absorbent coatings for the solar collecting surfaces, and a thermal storage and heat exchanger unit.

Success in constructing the solar energy facility suggested by the Meinels would appear to meet the demand for a large and assured new source of electric power. However, at the present time the costs of the Meinel project would be much greater than the price of electric energy produced by current means. But they contend that the costs of solar-produced electricity will become attractive as the costs of current energy production rise due to the increasing scarcity of conventional fuels.

Regarding the adverse impact of solar energy collection upon the environment, the Meinels suggest their project would be particularly desirable relative to other energy sources for two reasons. Since there is no combustion there would obviously be no combustion products. And since there are no fission or fusion processes involved there would be no radiation leakage or danger of any type of nuclear explosion.

There are environmental liabilities associated with the solar collection facility, but the Meinels believe these to be much less than would be the case with alternative means of energy production. First, an area of about 490 square miles of desert in the American southwest would be required for the 1,000,000 megawatt facility. This is slightly less than five percent of the land area in only one southwestern state, Arizona. Given the extremely arid and uninhabited nature of the land which would be required, this is a small environmental price to pay for the amount of "clean" energy discussed by the Meinels. Nevertheless, some environmentalists may be expected to become concerned over this use of land which would undeniably be removed from use by the native flora and fauna. Second, considerable waste heat would need to be dissipated from the collecting facility. The Meinels plan would obtain water for cooling from the Gulf of California, and in the process convert saltwater to fresh. The remaining brine would be re-deposited in the Gulf. Their preliminary studies indicate there will be no adverse effect upon aquatic life by the slightly increased salinity of the discharge area.

An indirect environmental aspect of solar collection in the American southwest could, but would not necessarily, result from the need to transport most of the resulting electrical energy to other areas where the demand for it will exist, i.e., the west coast, mid-west, and eastern seaboard. This would mean that electric lines would fan out from the solar collecting facility which would, if above ground, require considerable land over which they would be placed, causing in some instances "esthetic pollution." Of course there is the possibility that progress in cryogenics will advance to the point where underground transmission to far points will be a real alternative to above-the-ground lines. In fact, very long transmission may require underground super-cooled cables to avoid energy loss.

At this point it is difficult to discern direct military applications for a solar collection facility.

Perhaps the most likely military application of the energy from a solar collection facility would be in a laser beam defense system designed to destroy or make inoperative attacking aircraft, incoming missile warheads, or satellites (including sensor satellites). The implications of such a system upon nuclear stability are beyond the scope of this discussion, as are the implications for the idea of exchanging defense emphasis for the current offensive emphasis.

Technically speaking it might be possible to construct in orbit a solar collection facility whose energy could be used in an offensive fashion, i.e., as an energy beam directed downward at earth targets. In the constructing stages such a device could hardly be hidden and would be extremely vulnerable, however, and thus it does not appear as a particularly useful offensive system.

The impact of an American solar energy collection facility upon U.S. balance of payments and world trade is not yet clear. One possibility ought to be mentioned, however. An important area of the world in which increasing energy needs are found is Europe which does not possess a climate favorable to the development of solar collection on the earth's surface. The closest suitable area for solar collection at ground level is the desert area occupied by the Arab nations. Thus it can be seen that development of the Meinel solar project might result in Europe coveting the Arab lands more for solar energy than oil, or in the Arab energy position being enhanced by having solar energy added to their very considerable reserves of oil and gas. In any case international cooperation and understanding may be required in substantial amounts if unpleasantries are to be avoided between the Arab states and others. Solar collecting facilities for Asia could also pose both hazards and opportunities for international cooperation.

Regarding funding for R and D on the solar facility they propose, the

Meinels write: "It is clear that the magnitude of the task and the probable length of time from initiation to development of a commercial product is too large for it to be economically feasible for implementation *by private industry or public utilities*. Utilities are already economically stressed to meet today's demand for power, and they *must* use interim methods—fossil and nuclear—regardless of the ecological problems involved. The importance to the people of the United States of a pollution-free source of electrical power and one that will not be depleted makes it appropriate that this project be initiated by funding from the Federal Government."

The interesting question then arises, if the government must commence and carry out the development of a solar collection facility, who then will own it, and how will it be regulated so as to operate in a fashion consistent with the current definition of the public interest?

The Meinels envision research monies being utilized by a group drawn from the university sector, industry, and the public utilities. Eventually, since the facility will cross the border between Mexico and the U.S. funding may become multinational in scope.

Conclusion

The author is willing to argue that while the more power/better environment problem is important it will probably be resolved in some fashion tolerable to most of mankind by an arrangement featuring either the single or multiple use of fast breeder, fusion, and solar sources.

The more interesting questions, or if one is a pessimist the more difficult problems, will likely occur in regard to the impact the new energy sources will have upon property ownership and use, arms control and disarmament (war), and upon economic relations between nations.

Chapter X

A CONTRIBUTION TO SYSTEMS ANALYSIS OF
ENVIRONMENTAL CONTROL:
THE CASE OF THE GERMAN FEDERAL REPUBLIC

by *Peter-Jörg Jansen*

The object of the following paper is to show that in handling environmental problems it is first of all necessary to look for more convenient decision-making structures, allowing more effective cooperation between the scientific and policy-making sectors of society. In particular I will examine the problem in terms of the political system and environmental challenges of the German Federal Republic.[1] The analyses will be divided into three sections: 1) Towards a more systematic approach of the environmental challenge; 2) Tackling the problems, structures and processes of the environmental challenge; and 3) Analyzing the methods and criteria of the environmental challenge.

Though environmental problems are evidently physical, they originate in a social and political milieu. For this reason, an attempt will be made to indicate some basic characteristics of sociological structures, since a change of the social and political environment may in the long run contribute to solving the problems of technological development. This may be an unconventional contribution to the subject, in view of the complex nature of the problem. However, it seems to be an opportune time for discussion of these subjects.

Towards a more systematic approach of the environmental challenge

If the density of pedestrians on a pavement is low or has reached some medium value, the full variety of possible movements of the pedestrians with respect to the routes they may take is retained. If the density increases, the possibility of variety continues to increase also, but will often result in massive obstruction of the pedestrian flow unless patterns of behaviour are developed automatically. They do develop. The visible information about flows of movement on the pavement results in a self-control of the system. Areas with identical directions of movement are developed here and there

---

1. The ideas expressed in this paper are based on many discussions with Prof. Dr. W. Häfele and Mr. H. Zajone, to whom the author expresses his gratitude.

without there being any rules. However, it is cybernetics[2] which deals with the conditions permitting such self-control in very complex probabilistic systems.

A national economy may be understood as a complex probabilistic system. The interconnectivity of the elements of such a system is so high that any detailed control of the system is bound to fail. However, at the same time, intensive communication among the elements permits the establishment of patterns of behaviour which guarantee a stability of the system. Declaring such patterns of behaviour to be rules established by convention, limits the possible variety of the system. This applies likewise to rules introduced by law, e.g., traffic rules. It is not possible to have cars running in any direction like pedestrians and then wait for patterns of behaviour to establish themselves. For cybernetics, the requirement of having to introduce rules is ultimately due to a lack of communication among the elements of the system. Therefore, one of the main concerns of cybernetics is to create possibilities for communication, thus avoiding repressive rules. According to the cyberneticists, positive and negative feed-backs and homostasis scanning possibilities of reaction in a probabalistic way are the first tentative approaches to new structures in a self-controlling system. In a free interpretation of this concept, structural proposals will be discussed which ascribe similar tasks to groups for problem analysis, think-capacities, and to responsible authorities.

Initially, this concept seems to be very close to the trial and error principle which obviously produced reasonable systems in nature. However, there is no need for explaining that the human society of today can no longer afford to proceed on the basis of this principle. One of the advantages of human society is its ability to use intelligence to keep the variety of a system within reasonable limits by foresight and selection. In this, we have reached the area of management, of politics whose task is the establishment of objectives but not detailed implementation. Once the command values are given and have been accepted, suitable structures will permit a self-control of the system. To that extent the objectives of political leaders are clearcut: fixing goals and maintaining sufficient variety. However, the actual implementation of this objective is very exacting and will be dealt with below.

Primarily, self-control functions excellently within the energy-producing nations. Although electricity is something which must satisfy the needs of the consumer without any delay, the meeting of these requirements in the FRG has been ideally adapted to those needs until recently. This is all the more surprising as there is a multitude of utilities. Up to a certain threshold of

2. Cf. Stafford Beer, *Cybernetics and Management*. (London: The English Universities Press, Ltd., 1959).

expenditure, these enterprises independently try to insure the continuity of fuel supply, which is based on a multitude of mechanisms within industries and among nations. So, in the way the abstract quantity of money first enabled exchange processes to be carried out to the degree of complexity evident today and thus contributed substantially to the possibility of self-control, a qualitatively new level must be introduced to enable exchange processes of a higher order. For instance, the development of nuclear power stations of the fast breeder type cannot be carried out by industry alone. Economical electricity and certain continuity of energy reserves are factors which make development of this type of plant desirable for the national economy. It is essential that a political structure be responsible for recognizing the necessity of such developments and then, with careful long-range planning, to initiate and support them. The national government is the logical choice here. Below we will discuss specific methods by which such planning could be implemented.

Whenever a new invention or technological development with broad social implications emerges, a governing structure is necessary to evaluate future costs and benefits in the light of the nation's goal hierarchy. Investments into the social infrastructure are too far away from industry's economic mechanisms of self-control to occur automatically and therefore need control by government. It is probable that the combination of a self-controlling system with an apparatus responsible for foresight and selection corresponds to a subjection of cybernetics to the general systems theory. Systems theory tries to elaborate precisely those control variables, by means of systems analysis and preference synthesis, which are the guiding principles of an adaptable system. We hope to show that structures can be indicated for this purpose which safeguard the variety necessary for long-term self-control of society.

Our *first hypothesis* states that there are structures that must be developed to enable society as a very complex probabilistic system to develop respective patterns of behaviour on the basis of sufficient variety, i.e., a minimum of prescribed rules which allow actions to be taken in the direction of an agreed upon goal and a reaction to be generated when there are disturbing influences. Of course this implies a behaviour which painstakingly conforms to the existing system and might prevent any development and change of that system. Hence, everything depends on the way in which this elaboration of the goal hierarchy is designed. The plan formulated in the first hypothesis would ideally be designed for reacting to the changing goal hierarchies necessary for society.

In the cybernetic approach and also in the management science approach, which after all is one of the most important objectives of cybernetics, fixing of objectives is seen as largely detached from the system which is to implement those objectives. The approach of systems theory will provide a link between command values and the control system.

Hence, the *second hypothesis* to be formulated is: a system with sufficient variety with respect to the first hypothesis can be applied in the establishment of objectives to overcome conformity and yet provide for dynamic change of those objectives. The dilemma which could possibly arise here between a structure which determines the control variables and the ideal of a readily adaptable system is regarded as being solved because a system which has obtained a high degree of adaptability through intensive and flexible structures of communication will also question its own objectives. Because of the variety of communication directed to it, such a system is capable of establishing adaptable objectives. So, the second hypothesis is based on the conviction that the values applied to evaluate alternative objectives are subject to change by the same system for which these objectives are to be evaluated. Accordingly, "management" must be involved in the system's intense communications network. This implies that management is only one element of the system and that it is simply a matter of finding those structures of communication which will bring management into the self-control principle of the system.

This has brought us to a point of potential difficulty. Empirically it must be stated that the present structures do not allow full communication or discussion of objectives and attempt to thwart any developments with even short-range detrimental economic affects for an individual company. Obviously, there has never been sufficient thought about the consequences of laissez-faire economic growth to avoid the environmental problems which are of such concern to us today. Only a few individuals have challenged the institutions which have refused to recognize defects in our economic system or take steps to avoid undesirable economic, social, or environmental developments. Instead, the old system, whose variety is very limited, is retained and attempts to solve massive problems by isolated small steps.

Traditionally speaking, the implementation of the second hypothesis could be found in the fact that parliament allows a balance of interests and, by discussion, evolves a goal hierarchy for the benefit of society. It should also, however, provide for the establishment of structures which would guarantee the communication and variety of elements of the system necessary for self-control. This would necessitate long-range planning from which parliament escapes more and more as a consequence of the short periods of election. Instead, the administration takes up the role of long-range planner, seeking advice and information from representatives of science and industry. So far, the administrators have not been able to establish the intensive communication among the elements of the system which would keep the system alive. Indeed, the question must be raised whether the administration does not support structures which, to some extent, have resulted in developments negative to the good of society.

In an attempt to distribute power, the government is divided into three branches; legislative, executive, and judicial. At this time, however, it seems apparent that the executive branch is dominant and formulates most of the future plans. In the schema being developed here, it is the legislative branch which would determine national goal hierarchies. To assist the legislature in the evaluation of new programs, there would be an independent advisory agency whose primary function would be to submit reports on the broad implications of new projects and suggest possible alternatives. This would then allow the legislature to rationally discuss objectives and adopt only those ideas which would fit into the long-range goals of the country. Such an agency would guarantee a continuity of independent information by employing persons with a high degree of competence in social sciences, technology, and the physical sciences. In the past such agencies have been attempted, but they were not sufficiently problem oriented and received little recognition.

In the discussion of this advisory and analyzing structure, it is especially the aspect of a communicative society which must be taken into account. It is maintained that the structure to be created here justifies the second hypothesis and in this way, by its suggestions, also implements the first hypothesis. How this sector is to be established, permitting foresight and selection in establishing future national goals, will be described below.

Tackling the environmental challenge: Problems, structures, and the advisory process

Environmental problems themselves suggest many possible solutions. The importance of establishing social structures to avoid the creation of additional threats to the environment has already been indicated. The interrelationship of technical, biomedical, and social problems emerges immediately when a single ecological problem is considered.

Air pollution, for example, is a result of three major sources: the energy sector (power stations, domestic heating), the traffic sector (conventional cars, Diesel vehicles, airplanes), and industry. Technical measures to reduce this pollution require determination of output priorities and financial capabilities. Furthermore, there is the difficulty of determining the extent and nature of pollution sources, then assessing costs of specific corrective measures. In the example chosen to highlight this problem, an attempt will be made to compare the air-pollution caused by coal-fired power stations and light water reactors.[3]

3. P. Jansen, S. Jordan, and W. Schikarski, "An Approach to Compare Air Pollution of

To arrive at a comparable basis, we assume that 200 TWh of electrical energy would be generated alternatively by coal-fired power stations or nuclear power plant. This roughly corresponds to the amount of electricity generated by coal in the FRG in 1970. The quantities of pollutants emitted from the stacks of the coal-fired power stations under favourable technical conditions (with older power stations the pollution is even higher) included:

SO_2 1.8 . 10^6 t/a
dust   6 . 10^4 t/a
Ra   1.5   Ci

from boiling water reactors

Xe 133 4.5 . 10^8 Ci
Kr 85   4.5 . 10^4 Ci

from pressurized water reactors

Xe 133 6 . 10^4 Ci
Kr 85   6   Ci

from reprocessing plants
Kr 85   4 . 10^6 Ci

The problem is a comparison of these different materials in terms of their toxic effect. This initially requires data about their concentration in the air. We have left out the problem of conurbations, because only a relative comparison of the toxic effects will be possible. The resulting toxic concentrations also depend on the assumption that there are removal mechanisms for these toxic materials.

Since no valid data were available to us we did not assume any removal mechanisms in our first model and simply calculated the concentration produced on the basis of the annual emission. We have used three models of removal mechanisms and calculated the resulting equilibrium concentrations. This showed that the present uncertainties with respect to assumptions of residence times and removal rates have a much stronger influence on concentration than other assumptions in the model. It is, however, apparent that a total lack of consideration of removal mechanisms produces much different results. This can be explained very briefly:

The toxic effects of different fuels become comparable if the concen-

---

Fossil and Nuclear Power Plants," IAEA Symposium on Environmental Aspects of Nuclear Power Plants, New York, August, 1970.

trations are regarded in relationship to the maximum permissible concentrations. In doing so one assumes that the assessment of the maximum permissible concentration is based on a uniform philosophy.
Though this is not wholly valid, we nevertheless performed this analysis to come to a better understanding of the problem.

For coal-fired power stations, the ratio between actual and maximum permissible concentration is:

ratio: $\frac{\text{actual}}{\text{max. permissible}}$	without removal mechanisms	with removal mechanisms
SO_2	7.2	0.24
dust	1.2	0.06
Ra	$3 . 10^{-3}$	$1.5 . 10^{-4}$

Certainly, synergistic effects add an additional threat here. For the sake of experiment we first added only the toxic effects. In this case, the comparison among power stations looked like this:

ratio: $\frac{\text{actual}}{\text{max. permissible}}$	without removal mechanisms	with removal mechanisms
coal	8.4	0.30
boiling water reactors	0.02	0.005
pressurized water reactors	$2.7 . 10^{-6}$	$6 . 10^{-7}$
reprocessing plants	0.027	0.20

What can be seen from this, despite the simplifications?
1. If one leaves aside the removal mechanisms, which has often been done in previous discussions, coal will have a much poorer pollution record than nuclear energy.
2. If one looks only at nuclear power stations, leaving aside reprocessing, the frequently asserted opinion is correct that pressurized water reactors are better than boiling water reactors.
3. If one includes reprocessing plants, they dominate the toxic effects and in this way render the discussion between pressurized and boiling water reactors meaningless. This applies particularly if removal mechanisms are considered.
4. If one takes into account the removal mechanisms, it is not at all evident

that nuclear energy should be preferable to coal-fired power stations.

Hence, we might conclude that one should concentrate on the siting problem of reprocessing plants and the retention of Kr 85 in reprocessing. There are, however, other research efforts which appear to be even more important. These might include:

1. Provision of improved data on residence times and removal rates of toxic materials.
2. Investigation of the synergistic effects of toxic materials.
3. Determination of comparable standards of toxic materials.
4. Studies of the effectiveness of various removal mechanisms.
5. Elaboration of models of the distribution of toxic materials in the atmosphere.

Obviously, these problems require the collaboration of biologists, physicists, meteorologists, and so on. Once better data are available, it may perhaps be possible to draw more specific conclusions and give new impetus to research. From among the problems arising in this connection, only one will be singled out, namely the question of comparability or the determination of comparable standards of toxic materials.

The toxic materials which must be investigated within the framework of environmental research have very different properties. Besides studies of the mechanisms of their generation and their effects, comparison analyses must be carried out to prove what alternative strategies are suitable to improve environmental conditions and which toxic materials should receive the most immediate attention. These analyses are possible only if the toxic effects of different fuels can be compared. The problem of agreeing on a common philosophy for determining maximum permissible toxic concentrations depends heavily on the level at which toxic materials in the air actually become dangerous. So far, there has been no such common philosophy. Sufficient studies have not yet been conducted to determine what variables besides concentration (such as time of influence, synergistic effects, etc.) should be considered.

Guidelines must be worked out allowing specialists in medicine and biology to more closely define the maximum permissible conditions satisfying the requirements made above. It is also very important to agree on criteria and scales by which evaluation of the toxic effects will be made. If weak points in environmental protection are to be recognized and priorities established, it is necessary to think about criteria which can be used to evaluate different environmental influences. Only if such sets of criteria are made available can the results of environmental analyses be represented making it possible to evolve recommendations with respect to measures for protecting the environment. The establishment of these sets of criteria and the formal preparation of an evaluation of various environmental influences and counter measures has a major effect on social problems. It is not only new theories in economics, human ecology, and social medicine which must be worked out and introduced into a social process of discussion, but also a legal framework which would allow such theories to become effective in new social structures.

There is an obvious need here for an advisory committee or sector to help make the government more aware of the interrelatedness of socio-political and technical problems by discussing value preferences and then evaluating the problem at hand with scientific objectivity. Offering advice to decision-makers by presenting a small number of alternatives along with their conditions and consequences is a highly political affair. An organization which would appear to be best suited to handle such a function must seek to avoid hierarchical rigidity which would perpetuate administrative patterns of behaviour. Care must be taken in the selection of members to represent the various disciplines of science, social sciences, industry, administration and, very importantly, the general public in order to achieve a balance of interests and avoid strong lobbies. The competition between analytical groups and criticism by expert advisers representing various interests must also be taken into account to avoid falling victim to pretended arguments. Only if there is permanent communication between the politicians, engineers, and scientists on the one hand and the critical citizen on the other, will it be possible to speak of a communicative society which enables the process of self-control to be effective. A tentative approach seems to exist through a suitable staffing of an advisory body and the participation of many different analytical groups. Therefore, the main points of this advisory structure would be:
- the technically correct analysis of possible actions with an indication of the consequences and the criteria under which the respective alternatives would be brought to bear. This requires the establishment of analytical capacities.
- the interdisciplinary qualified staffing of the advisory body which would guarantee a disclosure of hidden preferences on the basis of the analyses and, by mutually constructive criticism, an elaboration of alternative solutions. Including the consequences they would be likely to entail, these could be used as aids in decisionmaking by political agencies.

The process outlined above still requires detailed consideration, both in terms of method and of structure, but would definitely aid in expanding a public consciousness in a participatory democracy.

In addition, initiating sciences[4] which have been very much in the background and whose formal character has been academic, need to be estab-

4. Jürgen Habermas, *Technik und Wissenschaft als "Ideologie,"* (Frankfurt: Suhrkamp, 1969).

lished. This applies to most areas of the social sciences, but also to research into the process of decision-making. We are good at establishing models and modes of procedure for individual problems but this produces solutions which maintain existing structures in the sense mentioned above. Models which would take into account the overall situation and establish inter-dependencies between different areas of society are largely absent.

Environmental problems, for instance, are of such concern to all of us and are so closely interwoven with economics, law, human ecology, and social medicine that there is an acute need to deal with them by means of more modern methods directed at a society of the future. These environmental problems cannot be coped with only by therapeutic measures but need a structural analysis dealing with the fundamental relations between cause and effect and resulting in the projection of a society in which problems of an environmental nature have been solved satisfactorily. Restricting these considerations, for instance, to environmental problems and energy gener-ation limit the variety of the social system.

A cursory examination of the problems surrounding energy production along the Rhine River is an excellent example of the close interrelationship of technical, environmental, and social problems. In a first approximation, 7 GWe of light water reactors can be installed on the Rhine with a maximum permissible differential temperature of 3°C (in comparison to normal temperatures) and a maximum temperature of 25°C if one wants to be able to guarantee 80% operation of these stations in different years without exceeding these limits. Is it economically worthwhile to permit 5°C and 28°C in view of the hazards arising from the synergistic effects of water pollution if this allows the installation of only another 5 GWe? Do we want to bio-logically sacrifice our rivers to the advantages of electricity generation or do we want a beautiful Rhine? What would be the costs to the national economy of a change to air cooling, and what is their relationship to the "costs" to society of a changed bioclimate along the Rhine?

This is only one of many far-reaching examples and is intended to justify why this paper does not deal with specific measures within the framework of the colloquium but instead pleads for thought about the structures nec-essary for a final elaboration of reasonable specific measures. Such struc-tural proposals include, also, the establishment of a center for environmental studies which would provide information on technological and economic trends and recommend suitable environmental actions. At the systems technique group at the Institute for Applied Reactor Physics of the Karlsruhe Nuclear Research Center we try to achieve some progress with practical analyses corresponding to the requirements shown above. We think we are not isolated in our general approaches and detailed analyses. However, intensive communication with all concerned sectors—political, scientific, sociological, industrial, and the general public—has yet to be established.

Analyzing the environmental challenge: methods and criteria

Finally, some thought must be given to the formal support for the advisory process outlined above. The analytical results with respect to alternative actions must be represented in such a way as to make them useful for a discussion within these advisory bodies. In view of the multitude of possible criteria fur judging the alternatives, one must systematize and rationalize the process of evaluation. This makes it desirable to find the most objective quantification of these terms in order to embark on a rational analysis. However, it should be taken into consideration that the process of quantification itself raises a number of problems and therefore the rational analysis thus carried out can provide only support, not act as a replacement for recognition of relationships and significant criteria. If quantitative methods are used, careful attention must be payed to the interpretation of the results. In evaluating alternatives qualitative arguments should also be taken into account. In any case, "requalification" as the opposite of quantification, is one of the main problems which must be solved to permit consideration of these evaluations within a general political context. This requalification is necessary because, due to the diversity of criteria applicable within one alternative, a numerical comparison between the criteria is impossible and only grades or categories of evaluations can be ascertained. This is the background on which all the evaluations proposed below should be seen.

On the basis of analyses it is possible only to arrange alternatives in a preferential sequence as far as their evaluation is concerned. However, in the actual determination of priorities it is important to see what weight is attached to the specific criteria; that is, the significance of the criteria must be harmonized. This should be done by an iterative process because it is only knowledge of the effects of alternatives which provides a clearer view of the importance of those criteria. Since the subjective concepts of the groups participating in decision making will be different, the process of establishing priorities should not be carried out as a seemingly exact mathematical process but as a clearly structured discussion. The efforts taken in evaluating systems analyses must pursue the goal of representing the contributions made by these projects to each different criterion to clearly indicate the causes for alternate preferences arising from subjective evaluations. This will then allow a pragmatic evaluation and adaptation of subjective opinions. Precisely for this reason is systems analysis so important in evaluating alternatives.

Below, a tentative approach is made to indicate criteria for evaluating alternatives which are not yet aimed at specific environmental problems but have been arranged along general lines.[5] Their specification with respect to environmental problems would require a detailed discussion. The remarks made below will therefore show only one approach and are not intended to provide solutions to specific problems.

Cataloguing criteria involves several categories which permit different approaches to the evaluation of alternatives. These sets of criteria include classification, effects, and properties.

Criteria of classification allow an arrangement of the alternatives in accordance with different goal areas of social activities. These may be regarded as areas of life and constitute the link to the goal hierarchy of politics which influences their significance. They may be considered as basic materials (food, raw materials), infrastructure (traffic and transport, information and communication, energy), and environment (health, social environment, ecology) and they indicate the direction in which alternative measures are intended. The effects reflect the consequences of specific alternative activities. To establish a comparison between very different efforts, a relatively crude scheme of evaluation is suggested with the four indicators: economic effects, social effects, political effects, and technical and scientific effects. Elaboration of these criteria necessary for comparison must start with the introduction of a scheme of evaluation and detailed specific analyses in each case. The criteria listed below for evaluating these effects should be regarded merely as examples.

a) Economic effects

Information on economic effects includes data on the influence of such specific efforts to supply basic materials as increases in energy generation capability, a change in the price of electricity, improvement of the energy supply by increased performance of fuel cells and batteries, an increase in mass production, or utilization of raw material in industrial processes (reducing the silver consumption in the photographic industries). Possibilities of increasing efficiency of the infrastructure could include more direct communications via videophones, processing information by data banks and reducing scanning times, dissemination of information through programmed learning, and reducing travel and delivery times. An improved environment might reduce the industrial absentee rate due to illness and even help reduce the number of traffic accidents. All these taken together have an effect on the growth rate

5. Ein methodologischer Beitrag zur Beurteilung von technologischen Forschungs- und Entwicklungsvorhaben unter dem Aspekt ihrer Förderung durch die öffentliche Hand. Erster Ergebnisbericht des ad hoc Ausschusses "Neue Technologien" des Beratenden Ausschusses für Forschungspolitik, Dez., 1970. (Unpublished.)

of the gross national product, the productivity of labor, an expanding international trade, and so on.

b) Social effects

The specific information under this group of effects will be very hard to quantify in most cases, but this should not prevent the respective efforts from being taken. Potential information of basic materials may include reducing the expenditures for food relative to the average income. A negative example in the area of infrastructure might reduce the freedom of the individual, perhaps by establishing a monopoly on information through introduction of a government data bank or standardizing education by programmed teaching. More positive examples would reduce the cost of individual travel and increase the percentage of highschool graduates. In the area of environment detailed information could help reduce the amount of airborn dust, exclude industrial pollution of rivers, reduce the cost of housing construction, increase life expectancy, and foster an increasing harmony with nature.

c) Political effects

This is intended mainly as a summary of the effects on foreign policy. Hence, initially the effects of alternatives on the economies and societies of other countries should be investigated in the same way as outlined under "a" and "b" for the FRG. Other factors to be evaluated include the prestige to be gained by specific projects (for instance through the number of scientists coming or returning to Germany for research purposes in connection with a project) or independence from other nations (construction of an isotope separation plant) or the degree of international collaboration, especially in the sector of basic science (perhaps expressed by the number of scientists exchanged).

d) Technical and scientific effects

These include effects whose influence on the target areas is only indirect or cannot yet be estimated. Thus, this requires information on enhancement of interdisciplinary research, training in the handling of complex systems, extension of scientific and technical knowledge, order of magnitude of the required R&D capacities, and the number of scientists.

Criteria for evaluation of alternatives are not only the effects explained above but also such properties characterizing the alternate activities as financial expenditure, material expenditure (required installations, energy requirement etc.), required manpower, and time expended. The following characteristics for evaluation may be important in some cases.

– Risk. This includes chances of success of the alternative, residual benefits if only a part of the program can be implemented, different approaches for reaching the desired goal, and accuracy of the data on the various expenditures.

– Maturity. This is the extent to which the basis of the alternative has already been worked on scientifically and the likelihood of the structures resulting in success of the proposal.

– Graduality. This provides a possibility for developing an alternative in stages. Thus a promising program which has inherent risks for success can be started on a trial basis. For instance, a small fraction of the total expenditure could be allocated to tackle the first phase of an alternative and clarify its actual chances for success and permit a better evaluation or even revision.

The result of the process of evaluation should be represented not in terms of numbers but arranged by categories of the contribution of the alternatives to the individual criteria in a very clear way in the form of a matrix in order to permit a dynamic process of discussion in which the evaluation of the criteria and its consequences will appear gradually. For a process of this type the psychological experience indicated below could be useful.

1. If alternatives A_i are to be evaluated by the criteria K_j, the human power of judgment ceases to be sufficient at $j = 1.2$ at several A_i, if required to indicate a preferential arrangement of A_i in one step.

2. An absolute evaluation a_{ij} of each A_i with respect to K_j seems to be impossible in view of the conditioned nature of human value concepts.

3. A relative evaluation b_{ij} seems to be possible for all A_i with a fixed j. However, a word of caution for overly extensive scales of evaluation is indicated. An assignment of A_i to a small number of categories of evaluation is preferable because a fixed arrangement will not be possible due to the uncertainties involved.

4. The mathematical calculation of the arrangement of preferences would require not only a statistical analysis of the confidence intervals but also the knowledge of the weights g_j of the individual criteria. However, this is precisely the question at issue for the different groups which have to make a decision. Therefore, a preferential arrangement of A_i can gradually emerge only out of the discussion of the effects of alternative evaluations; that is to say, if the factors b_{ij} are known and are made clear by the partners.

This can only be a very tedious process of coming to an agreement. However, this process can be supported by a systems technique, for instance by a sensitivity analysis of the criteria or by applying the Delphi-technique (a questioning action carried out over several rounds but interrupted by discussions).

Now it would be necessary to start elaborating criteria for evaluating alternative environmental measures. This, however, requires analyses which exceed the possibilities open to a contribution of this type of paper. Very different underlying concepts must be employed depending on whether environmental protection is viewed in its technological aspects or regarded as a function of social and economic regional planning.

Conclusions

We will not arrive at solutions of the complex relationships in the field of environmental control if we start planning in detail, but only if we create the structures in which foresight and selection are integrated into the process of self-control of a very complex probabilistic system society. It is hoped that the improvements indicated in the joint action of society, science, and politics and the approaches toward improvement of the transparency of evaluations and decision processes will facilitate that communication between the elements in society, necessary to permit self-control of the system in such a way as to prevent society from moving to a dead end. To handle environmental problems the politicians should establish an advisory body meeting the conditions mentioned in this paper in order to make effective the cooperation of science, policy, and the public. Moreover, they should foster analytical capacities for the analysis of the manifold necessities and implications of decisions touching environmental problems.

Chapter XI

EVALUATION AND OUTLOOK

by *S. L. Kwee*

Any attempt to fairly summarize the numerous considerations of the colloquium, much less to place them accurately in an integrative general framework, is obviously an undertaking of more than considerable difficulty. By focusing on the international problem of meeting future human needs for electrical energy while at the same time preserving a livable environment, the colloquium has raised many important questions of concern both to scientists and conservationists. Considerable data have been presented here, and more information is clearly available elsewhere. But it is just as clear that still additional research into energy and related ecological problems has yet to be conducted, and some primary research targets have already been recommended. Options have been formulated here and a few concrete practical measures have been suggested. Nonetheless, the general impression remains that the most crucial issues have merely been touched upon, that we are still a long way from genuine solutions. Real progress is uneven, and the precarious balance of limited, partial agreements and regulations can easily be upset by single acts of negligence or improvidence. We are aware of vast problems and grave dangers and we suspect that some of the strategies in use tend to aggravate the situation rather than alleviate it. What we need is not only a descriptive analysis of objective facts and processes but also a normative synthesis of subjective insights and experiences. These common insights and experiences should be made more conscious and explicit. At this crucial juncture, it might be useful to adduce a synoptic re-evaluation of the general problem and to assess the adequacy of proposals on the agenda and partial solutions already adopted. This concluding chapter is thus offered as a supplement and as a tentative guide for continuous discussions on "energy needs and environmental problems."

Problems like "energy and environment" are controversial, complicated and consequential. We try to tackle and solve them along three basic lines of approach:
by multi-disciplinary inquiry, analysis and research,
by multi-agency consultation, agreement and regulation, and
by multi-national cooperation, coordination and integration of both

scientific research and practical action.

We convene technologists and scientific experts from various disciplines, from representative posts in universities, industries, political institutions and governmental bodies, and from various countries and international committees and organizations. We ask them to examine distinctive problems, options, strategies, and prepare papers, surveys, prognoses, progress reports. Through a process of more or less rational and efficient discussions, inference, deduction, evaluation and explication, some more or less pertinent conclusions may be reached, priorities formulated, projects and programs suggested. These are fed back to scientific institutions for further analysis, verification, experimentation, correction and elaboration, and to industrial and governmental institutions for application and implementation in practical measures and regulations, adjustments of industrial and political programs.

The crucial role in this whole process is played by "scientists", by scientific and technological "experts". In the New Industrial State the potential might of science is greater than most scientists realize. Scientific knowledge is the root of technological and industrial innovation, of economic, social and cultural development. Science has become a social and political factor of utmost importance, an object of social concern, economic investment and political management and control. Scientists themselves are still very reluctant to acknowledge this fact, to organize science to match this task and to mobilize themselves accordingly.

There are serious contradictions within the realm of science. Representatives of the "two cultures" are unfamiliar with each other's interests and styles. There are endless debates among positivists and dialecticians concerning value-free science versus politically involved science. The world of science is in turmoil, if not in downright crisis. Its foundations are shaken and its aims are discredited. Scientific technology provides us with instruments for increasing wealth and power, but at ever higher costs and risks.

"Environment" has become a key-word symbol, arousing growing concern, feelings of guilt, and of failing responsibility. "Energy", on the other hand, symbolizes the mysterious power and driving force of unbridled "development", of hypertrophic economic, industrial, and military growth. Our inability to solve such a seemingly innocent problem like "electrical energy needs and environment" might in last instance be due to the much graver problem of the insufficiency of science. The reach of science in human life and society is extensive, but the role of scientific rationalism is limited and partial. Scientific rationalism functions on tiny islands within an ocean of irrationality. Scientific objectivism is extravert and veils the subjective workings of science itself.

Private science and public science

"Big science" has become too vast in scope and range, too complex in its organization and management for individual scientists to cope with it. Two new lines of development result. One is the gradual emergence of generalized disciplines like cybernetics, systems theory, computer-aided information processing and forecasting. Undoubtedly these new disciplines will influence and change our traditional ways of thinking, reasoning, argumentation, and decision making. The second line of development concerns the overall problem of the social functioning of science. As the reach of science in society increases, more and more scientists are called to fulfil posts in social and political life, for which their purely academic training is hopelessly inadequate. The insights and abilities necessary for the performance of the new tasks have to be acquired on the basis of practical experience. This realm of "public science" is rapidly expanding, quite distinct in scope and purpose and requiring other assets and attitudes than "private science."[1] The methods of inquiry and analysis, verification and experimentation in private science differ from the art of reasoning and argumentation, debating and decision making in public science. But public science is also different from industrial and military application of scientific knowledge.

There are two factors promoting and accelerating the growth of public science: the necessity of science policy and the political conscience of alarmed scientists. Science policy involves scientists in social and political affairs. The management and planning of science is a case for professionals. This professionalism must be based on both political expertness as well as scientific understanding. Once beyond the ivory tower public scientists become more familiar and conversant with social and political matters and also become more sensitive to the political importance and social prestige of science. This sensitivity makes them also aware of the social implications and political consequences of scientific discoveries and technological inventions. Some "disastrous applications" of scientific knowledge and "disruptive secondary effects" of originally neutral technologies are so conspicuous as to arouse public alarm. Scientists blame themselves for not discerning in time these detrimental effects of their work and try to compensate for them by heightened political alertness and intentional searches for constructive and peaceful applications. This itself is a good thing, but good intentions are mostly ineffective because they still fail to intervene and take adequate actions. There are two reasons for this relative inadequacy. One is a methodical error

---

1. Sir Solly Zuckerman, *Beyond the Ivory Tower: The Frontiers of Public and Private Science* (London: Weidenfeld & Nicholson, 1970); cf. C.P. Snow, *Public Affairs* (London: Macmillan, 1971).

and the other is insufficient organization.

Scientists, like other people are not visionaries. They might try to devise ingenious technologies of information processing and forecasting, but even then they can at most foretell only possibilities. What will actually happen depends on numerous processes of collective design and decision. At the moment a scientific discovery or a technological invention is made, the possible impact on future situations can be estimated. But what the actual course of action will be, has still to be deliberately planned, organized and effectuated. All phases of research and development, application and realization have to be actually carried out, deliberately and intentionally. All these phases are subject to human intervention, design, purpose, and voluntary action. Simple scientific knowledge of "how to split the atom" does not include detailed knowledge of "potentially" disastrous derivations like the atomic bomb nor beneficial applications like the nuclear reactor and medical use of isotopes. Between the initial discovery and all later elaborations are numerous processes of social choice and political decision, explication of purpose and conscientious design, resolute effort, and concentrated labour. Scientists and technologists can and must share in these processes of social deliberation and collective actualization. They are committing an error by looking for emendations in the realm of private science instead of improving the organization and effectiveness of public science.

The second reason why attempts by scientists to effectuate more or less radical social and political changes have been relatively inadequate so far, is that these attempts are limited in practical scope and insufficient in implementation. Much has happened during the last three decades which deserves to be reconsidered in connection with our problem of "energy and environment." A new characteristic element is that the motivation and initiative for political change arose explicitly among scientists. Some of the most politically active and perhaps most influential scientists have participated in the so-called Pugwash movement. The history of this movement is illustrative for the original resoluteness and deliberateness of its participants and the gradual deflection of their purpose and strategy.[2]

The Pugwash movement owed much of its early momentum to the decisive shock and resulting alarm in the aftermath of Hiroshima. Nuclear physicists were the first to fully realize the destructive power and implications of atomic fission. Indeed the development of the bomb was made possible not only by the invention of ingenious means to concentrate and release atomic energy but also by the organization of industrial means, social effort, and political power —importantly including multi-national teams of atomic scientists—under

2. Joseph Rotblat, *Pugwash: A History of the Conferences on Science and World Affairs* (Czechoslovakian Academy of Sciences, 1967).

pressure of war. After the war, with a short interruption, this massive endeavour was resumed with still more eagerness, under pressures of the cold war. Some physicists repented their collaboration and raised an alarm. The *Bulletin of Atomic Scientists* became a medium of political awakening. A nuclear group tried to mobilize the world of science into protest and action. Their primary aim, as proclaimed in the Russell-Einstein manifesto of July 1955 and the Vienna Declaration of September 1958, was "complete disarmament and world peace." But at the 10th Conference on Science and World Affairs in London, September 1962, the members had to admit that the arms race had not been stopped, that complete disarmament had proven unrealizable. Their goal was modified and they pressed now for arms control or arms limitation, and even this was continuously frustrated by the development of new weapon systems, biological and chemical weapons, strategies of deterrence and the danger of nuclear proliferation. There was a curious shift in terminology, from "defence" to "deterrence," from "war" to "cold war," from a short-term strategy of disarmament to a long-term "peace strategy" on four frontiers: from competition in armament to cooperation in peaceful uses of scientific knowledge, from disarmament to development, from the military and political conflicts between East and West to the economic and social contradictions between North and South, and from a "techno-scientific" approach to a "technosocial" approach. This broadening of strategy obviously weakened the original impetus. In later years more attention was paid to the resolution of restricted problems at regional levels, which proved to be more successful. The Pugwash scientists have gained general recognition and social prestige but little political influence.

Big science in astronomical dimensions

In 1955, even before the Pugwash conferences actually started, a program of Atoms for Peace was launched, met worldwide approval and resulted in the foundation of the International Atomic Energy Agency. That the big atomic powers could invest large sums in peaceful nuclear research was undoubtedly due to the awakening of a social conscience. These sums, however, remain quite small compared with ever increasing military expenditures. Also, the boundaries between peaceful research and military application are not always easy to trace. Science is often all too willing to be moulded into powerful means of industrial and military development. The distinction between peaceful and military applications of scientific knowledge is to some extent artificial and sometimes a convenient political device to divert funds into interconnected research. "Atoms for Peace" was a ready pretext to redouble high energy nuclear research, now purely in the realm of private and applied science. Big Science entered a boom and developed into Megatechnics and Gigatechnics with a powerful and instrumental infrastructure and economic establishment. After 1957 the arms race was partly overshadowed by a more spectacular competition in space flight. This transition to extraterrestrial dimensions was not at all surprising, since in nuclear physics and technology the sally into extraterrestrial forms of energy was begun with the first realization of nuclear fusion in the hydrogen bomb. Are we on the verge of a journey to the planets and beyond? This gigantic and audacious adventure of mankind is having its preparatory phase in the mobilization of all too aggressive social and political energy on earth.

Those scientists who proposed to substitute "development" for "disarmament" did not realize that "development" would become a magic formula, generating competition among nations even more irresponsible than war. What we have witnessed in recent decades is an explosion in development, as catastrophic perhaps as open military conflict, because it is no less energy devouring and environment polluting. The surprising economic and industrial recovery of Japan and Germany, the economic booms after the formal but not real resolution of local conflicts, the unabating strife and competition between socio-economic and political systems, are aspects of one and the same development explosion. Political and military contradictions are not really solved, they serve as an unexpected, unintended, but nevertheless readily welcomed and exploited stimulus for uninterrupted development. The affluent and acquisitive society has become "growth addicted." And growth is misunderstood here, not in the sense of organic growth, but as economic and industrial expansion. Development has become compulsive, both in the so-called underdeveloped as well as in the highly developed countries.

Dennis Gabor, in his remarkable study "*Innovations: Scientific, Technological and Social*," writes: "At present there is a terrifying imbalance in innovations. After many centuries in which innovation was almost imperceptible, and after a few in which all technical progress was identified with human progress, we have now reached a stage in which innovation has been compulsive—but only technological innovation. A large vested interest has been created, apart from the military-industrial complex, embodied in the avant-garde industries and research organizations, which believes that it must "innovate or die". Man's landing on the Moon is the epitome of this development; a splendid triumph of applied science and of a brilliant cooperative organization at a time when most thinking Americans are filled with grave doubts about the sanity of their society."[3]

3. Dennis Gabor, *Innovations: Scientific, Technological, and Social* (Oxford University Press, 1970), p. 2.

Space flight and high-energy physics are the most ambitious and at the same time most expensive projects of human technology. That they are disproportionately so is not a mere coincidence—both have their origin in the same illusive mentality. Gigantomania has almost become a collective obsession. Huge material and social investments in these costly projects are not justified by the relatively poor scientific and practical returns,[4] but inspired by motives of big power status and prestige. Both scientists and the public at large are victims of a curious illusion that these projects serve "peaceful competition." This illusion sprang from feelings of repentance among physicists for their "scientific sin" of inventing the atomic bomb. The overcompensation of this repentance, meanwhile, has acquired astronomical dimensions, in Giga electron Volts and in billions of dollars, and is reaching byond the earth. This incommensurability of scope is the main source of disruptive effects. Man's astral aspirations make him oblivious of his terrestrial surroundings. Gigatechnological violence easily upsets the intricate ecological equilibrium, the checks and balances of man's biological and social environment.

Nuclear physicists no longer stand alone in the process of developmental escalation. Once explosion has set in, the means to cope with it are also inflating. Physicians try to compensate the secondary effects of their medical care by curbing the population explosion through contraception. Agriculturists, all too eager to secure more food for the teeming millions by biochemical means, have to deplore a "silent spring." Mechanical, chemical and other technologists, in their attempts to create welfare, comfort, full employment, wealth and luxury in overcrowded habitats, desperately have to cope with stench, filth and waste, with air, water and soil pollution, with noise, jam, psychic stress, and other forms of "technological injury."[5]

Obviously, it is unfair to saddle only electric energy agencies with problems of environment. They are not solely er even mainly responsible for the most serious problems. The issue is much bigger and more complex. Of course, the possible contributions by energy technologists are various and numerous. They can match the requirements formulated by biologists, ecologists and economists: finding technological means for controlling the environmental effects of electric energy production, finding substitutes for electricity consumption, finding and improving production technologies with a more economic conversion rate and other source materials. But the question can

4. The value of returns of Big Science in general and of high-energy physics in particular has been increasingly questioned. See Philip W. Anderson, "Are the Big Machines Necessary?" *New Scientist*, LI (2 September, 1971), 510–513; Lewis Mumford, *The Pentagon of Power* (London: Secker, 1970) is still more eloquent, especially chapters 9–11.

5. See J. Rose, ed., *Technological Injury: The Effect of Technological Advances on Environment, Life, and Society* (London: Gordon & Breach, 1969).

also be turned the other way round by criticizing the steady increase in energy demand. The problem of energy should not be discussed as such but rather within the context of the general development explosion. It is impossible to tackle the problem if we only talk about reducing the growth of electricity consumption; it is a question of total energy consumption. If we want to change the total energy demand the whole pattern of our way of living must be changed and particularly the continuing contradictions between socio-economic and political systems must be reconsidered. The concern for environment points to changing perceptions and changing values in modern societies. These changing values should also be taken into account in judging the impacts of energy on environment. If the growing energy demand depends on a policy of full employment by a continuous expanse of industries, limitation of energy production and consumption would have undesired economic effects. On the other hand limitation of energy production would be a powerful medium of development control. Moreover, our conception of growth might be too narrow, taking into account only economic growth and being insufficiently aware of social and human growth which might ultimately involve the necessity of limiting economic growth.

All these and other problems must be discussed within a more integral perspective. Instead of designing programs which are apt to tackle symptoms rather than causes, we could then better understand the underlying process and free ourselves first of all from the obsessive development compulsion. Our most ingenious technologies are based upon physical, chemical, and biological realities already existent in inorganic and organic nature. We devise new technological systems, which subsequently must fit within the overall system of human ecology. It is absolutely necessary that we constantly keep in mind this fundamental background of our life and society. This ultimate totality should be the basic framework in which our scientific judgments, technological realizations, social choices, and political decisions take shape. The most inclusive conception or model of the world we live in, derived from a total synthesis of our most recent scientific insights and experiences, is that of "Spaceship Earth." Within this self-contained open system all transformations of matter and energy, homeostatic processes of interaction of self-regulating subsystems, metabolic and ecological chains, biological evolution and socio-cultural history are included.

The system is self-contained in a material respect, since after the separation of the Earth (virtually) no material exchange with the outside world has taken place and all material components—physical and chemical matter, geological and mineralogical formations—upon which all processes of evolution depend—chemical, biochemical, biological, and cultural evolution—are contained in it. The system is open in energetic respect in so far as exchange of radiation energy takes place—solar energy and cosmic rays, heat radiation from the earth—and in so far as the Earth is involved in planetary and satellite movement-energy of tides, and diurnal, lunar and seasonal rhythms, important in biological and cultural processes. During the four and a half billion odd years of separate existence of Spaceship Earth, evolutionary subsystems developed in mutual multi-loop feedback regulation. Evolutionary lines originated from newly emergent structures of intersystematic order: autocatalysis, cell structure, multicellular organism, memory and learning, conceptualization, philosophy and science as strategies of mind, tool-design and instrumentalization, social structure and institutionalization. In a relatively short span of time, Man succeeded in creating substitutional systems of material and cultural design for the fulfillment of his needs, wants, desires, and whims. These systems function on a technological infrastructure of material and energetic transformations. To this end Man has preyed upon all kinds of more or less available sources of matter and energy. By the substitution of new systems the intricate internal relations of Spaceship Earth are seriously disturbed and disrupted. This intrasystematic disruption becomes manifest in "environmental problems" or as "technological injury."

Within this framework of reference a re-evaluation of the energy-environment problem may be attempted. In spite of all science fiction, Spaceship Earth will remain our sole and total habitat for many generations to come. As far as we depend on science for the necessary tools and strategies of development and survival, we have to reorganize science in the light of this all-inclusive task. We know that Big Science has become a real power. The extension of this power is very uneven in the realms of private science, applied science and public science. The "high energy" and "dynamic development" aspects have been overemphasized. The intricate interrelationships and intra-systematic structures are still scientifically underdeveloped. "Economics" and "ecology" must be coordinated and perhaps integrated into a new discipline of "ecoumenology." All problems of energy sources, energy production and consumption, procedures and technologies have to be evaluated within a global perspective. When fundamental and integral solutions cannot be achieved at the moment, shortterm and partial solutions are only admissible in so far as they keep the chances open for more radical solutions afterwards. "Energy experts" and "environment experts" have to cooperate both in research and development in private science and applied science as well in political realization in public science.

Quoting Glenn Seaborg: "Our evolving mankind must learn to exist as an integral and contributing part of the earth that up to now has supported it unquestioningly. This can be achieved only by the formulation and application of a whole new scientific outlook and new ecological-technological relationship. This relationship must be based on a non-exploitive, closed-cycle way of life that is difficult to conceive of in terms of the way we live today. We will have to think and operate in terms of tremendous efficiencies. We will have to work with natural resources, energy, and the dynamics of the biosphere as a single system, nurturing and replenishing nature as she supports and sustains us."[6]

We have to keep in mind more carefully and accurately the limiting conditions within which all future strategies will have to operate. Those strategies which allow us only decades of breathing-time have to be rejected. Suppose that research in plasma physics opens new vistas of fusion energy technologies to be realized within four or five decades; are we free from the obligation to economize our use of fossil fuel resources in the meantime? Even when MHD research gives promising results of higher conversion rates, we must not forget that the total amount of energy transformation involved in steadily increasing energy production and consumption in the long run will have a devastating influence on climatic conditions, irregardless of whether we dispose of waste heat by cooling or by radiation. The increase in energy consumption can be interpreted also as an indication of increasing social and psychic unrest. And psychic unrest in its turn is related to the instability of social systems.

Conceiving of science as man's most important cybernetic medium in steering Spaceship Earth, how can we evaluate the role of science not only as to its impact upon the earth, but also its meaning towards man himself? Science has contributed to the explication and extension of man's world-picture (private science) and to the reshaping of human environment (applied science), but it still has failed to account rationally and efficiently for the effects of man's actions and designs (public science). There are attempts to supplement, augment, extend, and improve man's sensory and motor capacities and even his genetic inheritance, and to modify him by directly influencing his bio-psychic functions,[7] but these attempts are too narrow in scope and hence quite insufficient to magnify and improve his cybernetic qualities. For a real solution of energy-environment problems we urgently need, apart from technological innovations, science-based social and political alternatives.

6. Glenn Seaborg, "Environment and Society in Transition: Scientific Development, Social Consequences, Policy Implications," *Annals of the New York Academy of Sciences*, CLXXXIV (7 June, 1971), 685.

7. Cf. David Fishlock, *Man Modified: An Exploration of the Man/Machine Relationship* (New York: Funk & Wagnalls, 1969). The biotechnic approach to man relies more on "microminiturization" than on "big science."

Interests and concerns, needs and values

Our involvement in problems of environment reflects a growing concern and solicitude. The secondary or external effects of our technologies of energy production and consumption compel us to reconsider our economic and social system in a more integral perspective. It is no longer admissible to accept the continuously increasing energy demand as given. Economic calculations of costs and benefits, investments and returns are, from a broader point of view, inadequate and unjust. The direct benefits and returns tend to accrue to the happy few, while the neglected social costs and the necessary additional investments most often become a public responsibility.

Science and technology are objects of social and economic investments and, as a result, scientists and technologists have become the subjects of others' vested interests. Scientific education is a process of upgrading. Investments in scientific education and technological research are motivated by the expectation of returns in terms of scientific discoveries and technological inventions. But until fairly recently inventors served rather primitive needs and human desires. Research planning is directed by economic and social interests and guided by vague ideas about "fundamental importance." Only by probing into the integral relationship and dynamic equilibrium of natural and social systems have scientists become aware of the rapidly deteriorating state of the world in which they live. Quoting Dennis Gabor once more, "The most important and urgent problems of the technology of today are no longer the satisfaction of primary needs or of archetypal wishes, but the reparation of the evils and damages wrought by the technology of yesterday. We cannot stop inventing because we are riding a tiger. Fossil fuels are threatened by exhaustion; so we must have nuclear power. Death control has upset the balance of population; so we must have the pill. Mechanization, rationalization, and automation have upset the balance of employment; what is it we must have? For the time being we have nothing better than Parkinson's Law and restrictive practices."[8]

Because of the fragmentation of specialized science and divided loyalties of scientists in contradictory interest groups, most scientists are still only dimly aware of the rapidly growing imbalance or even exploding character of the present situation. When multi-disciplinary, multi-agency, multi-national confrontations and deliberations do not only result in objective delineations of problem territories and further research programs, but also in evoking a dramatic understanding of what is actually happening, feelings of concern

8. Gabor, p. 9; cf. Dennis Gabor, "Science, Growth, and Society: A New Perspective," *OECD*, 1971.

may be translated into explicit purpose and solid organization. The predominant organization of private and applied science on the basis of internally divided and mutually contradictory interests should be systematically transformed into a self-organization of public science based on integral understanding and universal concern, conscientious integrity, and methodical rationality.

The dramatic character of "world dynamics" is impressively demonstrated in the computer-model designed by Jay W. Forrester and Dennis Meadows of M.I.T. on request of the so-called Club of Rome.[9] According to Forrester, social systems behave contrarily to what we would intuitively expect. Strategies designed on the basis of a purely intuitive model are apt "to address symptoms rather than causes" and tend to be inadequate in three ways: a) by attempting to operate through points in the system that give little leverage for change, b) by applying a policy alleviating the situation in the short run but aggravating it in the long run and eventually causing deepening difficulties after a sequence of short-term actions, and c) by creating many unexpected side effects so that suppressing one symptom only causes trouble to burst forth at many other points. Forrester's dynamic computer-model shows the interactions between the five factors considered most important by the Club of Rome: world population, industrialization, depletion of natural resources, food production, and pollution.

The graphs and charts produced by Forrester show routes, interconnected by "multi-loop nonlinear feedback systems," into possible world situations, carrying us along or through abysses and natural disasters. The general impression so far is that we are in fact left with little choice. On the basis of the model, an "optimal strategy" can be calculated by which disasters would be minimal. The general outlook seems to be rather gloomy. Moreover, the computer-model is only a technical aid for political decisions and social options. How these decisions should be carried out in practice is another unsolved problem.

We have two reasons for mentioning Forrester's research in connection with our problem of energy and environment. First, a computer-model designed along the same lines involving the factors "energy production," "energy consumption," "environmental pollution," and "science and engineering policy" would undoubtedly demonstrate that the issues which confront us are at least as complex and crucial as those studied in Forrester's model. Second, an important consideration, influencing all factors analyzed by Forrester but omitted by him, is science itself. Perhaps, the most essential point of leverage is science and engineering policy. By overcoming first of all

---

9. Jay W. Forrester, "Counterintuitive Behavior of Social Systems," *Technology Review*, LXXIII, No. 3 (January, 1971); *World Dynamics* (Cambridge: Wright-Allen, 1971).

the immaturities and developmental imbalances of science and technology, we may exert a decisive influence on all other fields of future development. Instead of directing scientific research and technological invention towards the fulfillment of primitive needs and archetypal desires and wishes, we may concentrate on analyzing and improving the processes within science that create emergent values. Our values are changed by the impact of science and technology,[10] and our perception of values is deeply influenced by our involvement in scientific endeavour. "Energy" is only an instrumental value, while "unpolluted environment" is an intrinsic value. Science and technology may be re-oriented by a conscious hierarchy of values and by stressing the priority of intrinsic over instrumental values.

Private science, applied science, and public science make scientists sensitive to different sets of values: objectivity and rationality, utility and power, social relevance and collective welfare. Scientific cooperation and teamwork require explicit acknowledgment of shared values and are apt to nurture and foster them. Traditional university training of scientists rarely recognizes the importance of value perception and value change. A deliberate place in the scientific curriculum for value training cannot be said to exist. The idea of "value-free" science is even contradictory to such a recognition. Yet, it is obvious that agreement, choice, and decision depend on confluence of values. Experimental studies of small groups clearly demonstrate the gradual emergence of "microcultures," patterns of collective behaviour and group cohesion upon which the success of group activities depend.[11] More studies of the subjective processes operating in smaller or larger groups of cooperating scientists pursuing various goals may reveal useful data. Social scientists have amply analyzed the development of subcultures and counter-cultures, but they seem to overlook similar developments among scientists themselves. Working as a scientist in private science, applied science, or public science, exclusively or in combination, is intimately connected with subjective developments in value perceptions and value hierarchies.

On the other hand, the growing concern for the environment clearly points to overall trends in value explicitness and value upgrading. Solution of the energy-environment problem implies the resolution of conflicting value priorities. The unconscious selfdeception of scientists, who, after first striving for complete disarmament, were gradually deflected towards arms control and peaceful development, might also be due to a confusion of

10. Cf. K. Baier and N. Rescher, eds., *Values and the Future: The Impact of Technological Change on American Values* (New York: Free Press, 1969); and D. Willner, ed., *Decisions, Values and Groups* (Oxford: Pergamon, 1960), particularly the introduction by Anatol Rapoport.

11. Edward Rose, "The Organization of Microcultures," in D. Willner, ed., *Decisions, Values and Groups*, pp. 171–175.

values. A critical re-evaluation of the relations and interactions of science, technology, and society with respect to "energy" and "environment," and a conscious reflection upon this re-evaluation process itself may contribute to the practical effectiveness of discussions and colloquia.

The subjective aspects of science are overemphasized here, because the problem of "energy and environment," although it can be studied separately, cannot be solved in isolation from the more inclusive and complex problem of the role of science in society. As an "objective" problem the question of energy and environment is hardly relevant. A normative synthesis of subjective insights and experiences will reveal the fundamental values, in terms of which the whole problem requires a new setting. Our failure or success to solve this problem radically will decide our future, not for some decades or even centuries, but for the complete survival of mankind.

ANNEXES

MESSAGE OF PRESIDENT NIXON TO THE CONGRESS OF THE
UNITED STATES. JUNE 4, 1971. OUTLINING A PROGRAM TO
ENSURE AN ADEQUATE SUPPLY OF CLEAN ENERGY FOR THE
YEARS AHEAD

For most of our history, a plentiful supply of energy is something the American people have taken very much for granted. In the past twenty years alone, we have been able to double our consumption of energy without exhausting the supply. But the assumption that sufficient energy will always be readily available has been brought sharply into question within the last year. The brownouts that have affected some areas of our country, the possible shortages of fuel that were threatened last fall, the sharp increases in certain fuel prices and our growing awareness of the environmental consequences of energy production have all demonstrated that we cannot take our energy supply for granted any longer.

A sufficient supply of clean energy is essential if we are to sustain healthy economic growth and improve the quality of our national life. I am therefore announcing today a broad range of actions to ensure an adequate supply of clean energy for the years ahead. Private industry, of course, will still play the major role in providing our energy, but government can do a great deal to help in meeting this challenge.

My program includes the following elements:

To Facilitate Research and Development for Clean Energy:

– A commitment to complete the successful demonstration of the liquid metal fast breeder reactor by 1980.
– More than twice as much Federal support for sulfur oxide control demonstration projects in Fiscal Year 1972.
– An expanded program to convert coal into a clean gaseous fuel.
– Support for a variety of other energy research projects in fields such as fusion power, magnetohydrodynamic power cycles, and underground electric transmission.

To Make Available the Energy Resources on Federal Lands:

– Acceleration of oil and gas lease sales on the Outer Continental Shelf, along with stringent controls to protect the environment.
– A leasing program to develop our vast oil shale resources, provided that

environmental questions can be satisfactorily resolved.
– Development of a geothermal leasing program beginning this fall.

To Assure a Timely Supply of Nuclear Fuels:

– Begin work to modernize and expand our uranium enrichment capacity.

To Use Our Energy More Wisely:

– A New Federal Housing Administration standard requiring additional insulation in new federally insured homes.
– Development and publication of additional information on how consumers can use energy more efficiently.
– Other efforts to encourage energy conservation.

To Balance Environmental and Energy Needs:

– A system of long-range open planning of electric power plant sites and transmission line routes with approval by a State or regional agency before construction.
– An incentive charge to reduce sulfur oxide emissions and to support further research.

To Organize Federal Efforts More Effectively:

– A single structure within the Department of Natural Resources uniting all important energy resource development programs.

THE NATURE OF THE CURRENT PROBLEM

A major cause of our recent energy problems has been the sharp increase in demand that began about 1967. For decades, energy consumption had generally grown at a slower rate than the national output of goods and services. But in the last four years it has been growing at a faster pace and forecasts of energy demand a decade from now have been undergoing significant upward revisions.

This accelerated growth in demand results partly from the fact that energy has been relatively inexpensive in this country. During the last decade, the prices of oil, coal, natural gas and electricity have increased at a much slower rate than consumer prices as a whole. Energy has been an attractive bargain in this country—and demand has responded accordingly.

In the years ahead, the needs of a growing economy will further stimulate this demand. And the new emphasis on environmental protection means that the demand for cleaner fuels will be especially acute. The primary cause of air pollution, for example, is the burning of fossil fuels in homes, in cars, in factories and in power plants. If we are to meet our new national air quality standards, it will be essential for us to use stack gas cleaning systems in our large power and other industrial plants and to use cleaner fuels in virtually all of our new residential, commercial and industrial facilities, and in some of our older facilities as well.

Together, these two factors–growing demand for energy and growing emphasis on cleaner fuels—will create an extraordinary pressure on our fuel supplies.

The task of providing sufficient clean energy is made especially difficult by the long lead times required to increase energy supply. To move from geological exploration to oil and gas well production now takes from 3 to 7 years. New coal mines typically require 3 to 5 years to reach the production stage and it takes 5 to 7 years to complete a large steam power plant. The development of the new technology required to minimize environmental damage can further delay the provision of additional energy. If we are to take full advantage of our enormous coal resources, for example, we will need mining systems that do not impair the health and safety of miners or degrade the landscape and combustion systems that do not emit harmful quantities of sulfur oxides, other noxious gases, and particulates into the atmosphere. But such systems may take several years to reach satisfactory performance. That is why our efforts to expand the supply of clean energy in America must immediately be stepped up.

1. Research and Development Goals for Clean Energy

Our past research in this critical field has produced many promising leads. Now we must move quickly to demonstrate the best of these new concepts on a commercial scale. Industry should play the major role in this area, but government can help by providing technical leadership and by sharing a portion of the risk for costly demonstration plants. The time has now come for government and industry to commit themselves to a joint effort to achieve commercial scale demonstrations in the most crucial and most promising clean energy development areas—the fast breeder reactor sulfur oxide control technology and coal gasification.

a. Sulfur Oxide Control Technology

A major bottleneck in our clean energy program is the fact that we cannot now burn coal or oil without discharging its sulfur content into the air. We need new technology which will make it possible to remove the sulfur before it is emitted to the air.

Working together, industry and government have developed a variety of approaches to this problem. However, the new air quality standards promulgated under the Clean Air Amendments of 1970 require an even more rapid development of a suitable range of stack gas cleaning techniques for removing sulfur oxides. In have therefore requested funds in my 1972 budget to permit the Environmental Protection Agency to devote an additional $15 million to this area, more than doubling the level of our previous efforts. This expansion means that a total of six different techniques can be demonstrated in partnership with industry during the next three or four years.

b. Nuclear Breeder Reactor

Our best hope today for meeting the Nation's growing demand for economical clean energy lies with the fast breeder reactor. Because of its highly efficient use of nuclear fuel, the breeder reactor could extend the life of our natural uranium fuel supply from decades to centuries, with far less impact on the environment than the power plants which are operating today.

For several years, the Atomic Energy Commission has placed the highest priority on developing the liquid metal fast breeder. Now this project is ready to move out of the laboratory and into the demonstration phase with a commercial size plant. But there still are major technical and financial obstacles to the construction of a demonstration plant of some 300 to 500 megawatts. I am therefore requesting an additional $27 million in Fiscal Year 1972 for the Atomic Energy Commission's liquid metal fast breeder reactor program—and for related technological and safety programs—so that the necessary engineering groundwork for demonstration plants can soon be laid.

What about the environmental impact of such plants? It is reassuring to know that the releases of radioactivity from current nuclear reactors are well within the national safety standards. Nevertheless, we will make every effort to see that these new breeder reactors emit even less radioactivity to the environment than the commercial light water reactors which are now in use.

I am therefore directing the Atomic Energy Commission to ensure that the new breeder plants be designed in a way which inherently prevents discharge to the environment from the plant's radioactive effluent systems. The Atomic Energy Commission should also take advantage of the increased efficiency of these breeder plants, designing them to minimize waste heat discharges. Thermal pollution from nuclear power plants can be materially reduced in the more efficient breeder reactors.

We have very high hopes that the breeder reactor will soon become a key element in the national fight against air and water pollution. In order further to inform the interested agencies and the public about the opportunities in this area, I have requested the early preparation and review by all appropriate agencies of a draft environmental impact statement for the breeder demonstration plant in accordance with Section 102 of the National Environmental Policy Act. This procedure will ensure compliance with all environmental quality standards before plant construction begins.

In a related area, it is also pertinent to observe that the safety record of civilian power reactors in this country is extraordinary in the history of technological advances. For more than a quarter century—since the first nuclear chain reaction took place—no member of the public has been injured by the failure of a reactor or by an accidental release of radioactivity. I am confident that this record can be maintained. The Atomic Energy Commission is giving top priority to safety considerations in the basic design of the breeder reactor and this design will also be subject to a thorough review by the independent Advisory Committee on Reactor Safeguards, which will publish the results of its investigation.

I believe it important to the Nation that the commercial demonstration of a breeder reactor be completed by 1980. To help achieve that goal, I am requesting an additional $50 million in Federal funds for the demonstration plant. We expect industry—the utilities and manufacturers—to contribute the major share of the plant's total cost, since they have a large and obvious stake in this new technology. But we also recognize that only if government and industry work closely together can we maximize our progress in this vital field and thus introduce a new era in the production of energy for the people of our land.

c. Coal Gasification

As we carry on our search for cleaner fuels, we think immediately of the cleanest fossil fuel—natural gas. But our reserves of natural gas are quite limited in comparison with our reserves of coal.

Fortunately, however, it is technically feasible to convert coal into a clean gas which can be transported through pipelines. The Department of the Interior has been working with the natural gas and coal industries on research to advance our coal gasification efforts and a number of possible methods for accomplishing this conversion are under development. A few, in fact, are now in the pilot plant stage.

We are determined to bring greater focus and urgency to this effort. We have therefore initiated a cooperative program with industry to expand the number of pilot plants, making it possible to test new methods more expeditiously so that the appropriate technology can soon be selected for a large-scale demonstration plant.

The Federal expenditure for this cooperative program will be expanded to $20 million a year. Industry has agreed to provide $10 million a year for this effort. In general, we expect that the Government will continue to finance the larger share of pilot plants and that industry will finance the larger share of the demonstration plants. But again, the important point is that both the Government and industry are now strongly committed to move ahead together as promptly as possible to make coal gasification a commercial reality.

d. Other Research and Development Efforts

The fast breeder reactor sulfur oxide controls and coal gasification represent our highest priority research and development projects in the clean energy field. But they are not our only efforts. Other ongoing projects include:

– Coal Mine Health and Safety Research. In response to a growing concern for the health and safety of the men who mine the Nation's coal and in accordance with the Federal Coal Mine Health and Safety Act of 1969, the Bureau of Mines research effort has been increased from a level of $2 million in Fiscal Year 1969 to $30 million in Fiscal Year 1972.

– Controlled Thermonuclear Fusion Research. For nearly two decades the Government has been funding a sizeable research effort designed to harness the almost limitless energy of nuclear fusion for peaceful purposes. Recent progress suggests that the scientific feasibility of such projects may be demonstrated in the 1970s and we have therefore requested an additional $2 million to supplement the budget in this field for Fiscal Year 1972. We hope that work in this promising area will continue to be expanded as scientific progress justifies larger scale programs.

– Coal Liquefaction. In addition to its coal gasification work, the Department of the Interior has underway a major pilot plant program directed toward converting coal into cleaner liquid fuels.

– Magnetohydrodynamic Power Cycles. MHD is a new and more efficient method of converting coal and other fossil fuels into electric energy by burning the fuel and passing the combustion products through a magnetic field at very high temperatures. In partnership with the electric power industry, we have been working to develop this new system of electric power generation.

– Underground Electric Transmission. Objections have been growing to the overhead placement of high voltage power lines, especially in areas of scenic beauty or near centers of population. Again in cooperation with industry, the Government is funding a research program to develop new and less expensive techniques for burying high voltage electric transmission lines.

– Nuclear Reactor Safety and Supporting Technology. The general research and development work for today's commercial nuclear reactors was completed several years ago, but we must continue to fund safety-related efforts in order to ensure the continuance of the excellent safety record in this field. An additional $3 million has recently been requested for this purpose to supplement the budget in Fiscal Year 1972.

– Advanced Reactor Concepts. The liquid metal fast breeder is the priority breeder reactor concept under development, but the Atomic Energy Com-

mission is also supporting limited alternate reactor programs involving gas cooled reactors, molten salt reactors and light water breeders.
– Solar Energy. The sun offers an almost unlimited supply of energy if we can learn to use it economically. The National Aeronautics and Space Administration and the National Science Foundation are currently re-examining their efforts in this area and we expect to give greater attention to solar energy in the future.

The key to meeting our twin goals of supplying adequate energy and protecting the environment in the decades ahead will be a balanced and imaginative research and development program. I have therefore asked my Science Adviser, with the cooperation of the Council on Environmental Quality and the interested agencies, to make a detailed assessment of all of the technological opportunities in this area and to recommend additional projects which should receive priority attention.

2. Making Available the Energy Resources of Federal Lands

Over half of our Nation's remaining oil and gas resources, about 40 percent of our coal and uranium, 80 percent of our oil shale, and some 60 percent of our geothermal energy sources are now located on Federal lands. Programs to make these resources available to meet the growing energy requirements of the Nation are therefore essential if shortages are to be averted. Through appropriate leasing programs, the Government should be able to recover the fair market value of these resources, while requiring developers to comply with requirements that will adequately protect the environment.

To supplement the efforts already underway to develop the fuel resources of the lower 48 States and Alaska, I am announcing today the following new programs:

a. Leasing on the Outer Continental Shelf—An Accelerated Program

The Outer Continental Shelf has proved to be a prolific source of oil and gas, but it has also been the source of troublesome oil spills in recent years. Our ability to tap the great potential of offshore areas has been seriously hampered by these environmental problems.

The Department of the Interior has significantly strengthened the environmental protection requirements controlling offshore drilling and we will continue to enforce these requirements very strictly. As a prerequisite to Federal lease sales, environmental assessments will be made in accordance with Section 102 of the National Environmental Policy Act of 1969.

Within these clear limits, we will accelerate our efforts to utilize this rich source of fuel. In order to expand productive possibilities as rapidly as possible, the accelerated program should include the sale of new leases not only in the highly productive Gulf of Mexico, but also some other promising areas. I am therefore directing the Secretary of the Interior to increase the offerings of oil and gas leases and to publish a schedule for lease offerings on the Outer Continental Shelf during the next five years, beginning with a general lease sale and a drainage sale this year.

b. Oil Shale—A Program for Orderly Development

At a time when we are facing possible energy shortages, it is reassuring to know that there exists in the United States an untapped shale oil resource containing some 600 billion barrels in high grade deposits. At current consumption rates, this resource represents 150 years supply. About 80 billion barrels of this shale oil are particularly rich and well situated for early development. This huge resource of very low sulfur oil is located in the Rocky Mountain area, primarily on Federal land.

At present there is no commercial production of shale oil. A mixture of problems—environmental, technical and economic—have combined to thwart past efforts at development.

I believe the time has come to begin the orderly formulation of a shale oil policy—not by any head-long rush toward development but rather by a well considered program in which both environmental protection and the recovery of a fair return to the Government are cardinal principles under which any leasing takes place. I am therefore requesting the Secretary of the Interior to expedite the development of an oil shale leasing program including the preparation of an environmental impact statement. If after reviewing this statement and comments he finds that environmental concerns can be satisfied, he shall then proceed with the detailed planning. This work would also involve the States of Wyoming, Colorado and Utah and the first test lease would be scheduled for next year.

c. Geothermal Energy

There is a vast quantity of heat stored in the earth itself. Where this energy source is close to the surface, as it is in the Western States, it can readily be tapped to generate electricity, to heat homes, and to meet other energy requirements. Again, this resource is located primarily on Federal lands.

Legislation enacted in recent months permits the Federal government, for the first time, to prepare for a leasing program in the field of geothermal energy. Classification of the lands involved is already underway in the Department of the Interior. I am requesting the Secretary of the Interior to expedite a final decision on whether the first competitive lease sale should be scheduled for this fall—taking into account, of course, his evaluation of the environmental impact statement.

3. Natural Gas Supply

For the past 25 years, natural gas has supplied much of the increase in the energy supply of the United States. Now this relatively clean form of energy is in even greater demand to help satisfy air quality standards. Our present supply of natural gas is limited, however, and we are beginning to face shortages which could intensify as we move to implement the air quality standards. Additional supplies of gas will therefore be one of our most urgent energy needs in the next few years.

Federal efforts to augment the available supplies of natural gas include:

- Accelerated leasing on Federal lands to speed discovery and development of new natural gas fields.
- Moving ahead with a demonstration project to gasify coal.
- Recent actions by the Federal Power Commission providing greater incentives for industry to increase its search for new sources of natural gas and to commit its discoveries to the interstate market.
- Facilitating imports of both natural and liquefied gas from Canada and from other nations.
- Progress in nuclear stimulation experiments which seek to produce natural gas from tight geologic formations which cannot presently be utilized in ways which are economically and environmentally acceptable.

This administration is keenly aware of the need to take every reasonable action to enlarge the supply of clean gaseous fuels. We intend to take such action and we expect to get good results.

4. Imports from Canada

Over the years, the United States and Canada have steadily increased their trade in energy. The United States exports some coal to Canada, but the major items of trade are oil and gas which are surplus to Canadian needs but which find a ready market in the United States.

The time has come to develop further this mutually advantageous trading relationship. The United States is therefore prepared to move promptly to permit Canadian crude oil to enter this country, free of any quantitative restraints, upon agreement as to measures needed to prevent citizens of both our countries from being subjected to oil shortages, or threats of shortages. We are ready to proceed with negotiations and we look to an early conclusion.

5. Timely Supplies of Nuclear Fuels

The Nation's nuclear fuel supply is in a state of transition. Military needs are now relatively small but civilian needs are growing rapidly and will be our dominant need for nuclear fuel in the future. With the exception of uranium enrichment, the nuclear energy industry is now in private hands.

I expect that private enterprise will eventually assume the responsibility for uranium enrichment as well, but in the meantime the Government must carry out its responsibility to ensure that our enrichment capacity expands at a rate consistent with expected demands.

There is currently no shortage of enriched uranium or enriching capacity. In fact, the Atomic Energy Commission has substantial stocks of enriched uranium which have already been produced for later use. However, plant expansions are required so that we can meet the growing demands for nuclear fuel in the late 1970s—both in the United States and in other nations for which this country is now the principal supplier.

The most economical means presently available for expanding our capacity in this field appears to be the modernization of existing gaseous diffusion plants at Oak Ridge, Tennessee; Portsmouth, Ohio; and Paducah, Kentucky—through a Cascade Improvement Program. This program will take a number of years to complete and we therefore believe that it is prudent to initiate the program at this time rather than run the risk of shortages at a later date. I am therefore releasing $16 million to start the Cascade Improvement Program in Fiscal Year 1972. The pace of the improvement program will be tailored to fit the demands for enriched uranium in the United States and in other countries.

6. Using Our Energy More Wisely

We need new sources of energy in this country, but we also need to use existing energy as efficiently as possible. I believe we can achieve the ends we desire—homes warm in winter and cool in summer, rapid transportation, plentiful energy for industrial production and home appliances—and still place less of a strain on our overtaxed resources.

Historically, we have converted fuels into electricity and have used other sources of energy with ever increasing efficiency. Recent data suggest, however, that this trend may be reversing—thus adding to the drain on available resources. We must get back on the road of increasing efficiency—both at the point of production and at the point of consumption, where the consumer himself can do a great deal to achieve considerable savings in his energy bills.

We believe that part of the answer lies in pricing energy on the basis of its full costs to society. One reason we use energy so lavishly today is that the price of energy does not include all of the social costs of producing it. The costs incurred in protecting the environment and the health and safety of workers, for example, are part of the real cost of producing energy—but they are not now all included in the price of the product. If they were added to that price, we could expect that some of the waste in the use of energy would be eliminated. At the same time, by expanding clean fuel supplies, we will be working to keep the overall cost of energy as low as possible.

It is also important that the individual consumer be fully aware of what his energy will cost if he buys a particular home or appliance. The efficiency of home heating or cooling systems and of other energy intensive equipment are determined by builders and manufacturers who may be concerned more with the initial cost of the equipment than with the operating costs which will come afterward. For example, better thermal insulation in a home or office building may save the consumer large sums in the long run—and conserve energy as well—but for the builder it merely represents an added expense.

To help meet one manifestation of this problem, I am directing the Secretary of Housing and Urban Development to issue revised standards for insulation applied in new federally insured homes. The new Federal Housing Administration standards will require sufficient insulation to reduce the maximum permissible heat loss by about one-third for a typical 1200 square foot home—and by even more for larger homes. It is estimated that the fuel savings which will result each year from the application of these new standards will, in an average climate, equal the cost of the additional insulation required.

While the Federal Government can take some actions to conserve energy through such regulations, the consumer who seeks the most for his energy dollar in the marketplace is the one who can have the most profound influence. I am therefore asking my Special Assistant for Consumer Affairs—in cooperation with industry and appropriate Government agencies—to gather and publish additional information in this field to help consumers focus on the operating costs as well as the initial cost of energy intensive equipment.

In addition, I would note that the Joint Board on Fuel Supply and Fuel Transport chaired by the Director of the Office of Emergency Preparedness is developing energy conservation measures for industry, government, and the general public to help reduce energy use in times of particular shortage and during pollution crises.

7. Power Plant Siting

If we are to meet growing demands for electricity in the years ahead, we cannot ignore the need for many new power plants. These plants and their associated transmission lines must be located and built so as to avoid major damage to the environment, but they must also be completed on time so as to avoid power shortages. These demands are difficult to reconcile—and often they are not reconciled well. In my judgment the lesson of the recent power shortages and of the continuing disputes over power plant siting and transmission line routes is that the existing institutions for making decisions in this area are not adequate for the job. In my Special Message to the Congress on the Environment last February, I proposed legislation which would help to alleviate these problems through longer range planning by the utilities and through the establishment of State or regional agencies to license new bulk power facilities prior to their construction.

Hearings are now being held by the Interstate and Foreign Commerce Committee of the House of Representatives concerning these proposals and other measures which would provide an open planning and decision-making capacity for dealing with these matters. Under the administration bill, long-range expansion plans would be presented by the utilities ten years before construction was scheduled to begin, individual alternative power plant sites would be identified five years ahead, and detailed design and location of specific plants and transmission lines would be considered two years in advance of construction. Public hearings would be held far enough ahead of construction so that they could influence the siting decision, helping to avoid environmental problems without causing undue construction delays. I urge the Congress to take prompt and favorable action on this important legislative proposal. At the same time steps will be taken to ensure that Federal licenses and permits are handled as expeditiously as possible.

8. The Role of the Sulfur Oxides Emissions Charge

In my environmental message last February I also proposed the establishment of a sulfur oxides emissions charge. The emissions charge would have the effect of building the cost of sulfur oxide pollution into the price of energy. It would also provide a strong economic incentive for achieving the necessary performance to meet sulfur oxide standards.

The funds generated by the emissions charge would be used by the Federal Government to expand its programs to improve environmental quality, with special emphasis on the development of adequate supplies of clean energy.

9. Government Reorganization—An Energy Administration

But new programs alone will not be enough. We must also consider how we can make these programs do what we intend them to do. One important way of fostering effective performance is to place responsibility for energy questions in a single agency which can execute and modify policies in a comprehensive and unified manner.

The Nation has been without an integrated energy policy in the past. One reason for this situation is that energy responsibilities are fragmented among several agencies. Often authority is divided according to types and uses of energy. Coal, for example, is handled in one place, nuclear energy in another—but responsibility for considering the impact of one on the other is not assigned to any single authority. Nor is there any single agency responsible for developing new energy sources such as solar energy or new conversion systems such as the fuel cell. New concerns—such as conserving our fossil fuels for non-fuel uses—cannot receive the thorough and thoughtful attention they deserve under present arrangements.

The reason for all these deficiencies is that each existing program was set up to meet a specific problem of the past. As a result, our present structure is not equipped to handle the relationships between these problems and the emergence of new concerns.

The need to remedy these problems becomes more pressing every day. For example, the energy industries presently account for some 20 percent of our investment in new plant and equipment. This means that inefficiencies resulting from uncoordinated government programs can be very costly to our economy. It is also true that energy sources are becoming increasingly interchangeable. Coal can be converted to gas, for example, and even to synthetic crude oil. If the Government is to perform adequately in the energy field, then it must act through an agency which has sufficient strength and breadth of responsibility.

Accordingly, I have proposed that all of our important Federal energy resource development programs be consolidated within the new Department of Natural Resources.

The single energy authority which would thus be created would be better able to clarify, express, and execute Federal energy policy than any unit in our present structure. The establishment of this new entity would provide a focal point where energy policy in the executive branch could be harmonized and rationalized.

One of the major advantages of consolidating energy responsibilities would be the broader scope and greater balance this would give to research and development work in the energy field. The Atomic Energy Commission, for instance, has been successful in its mission of advancing civilian nuclear power, but this field is now intimately interrelated with coal, oil and gas, and Federal electric power programs with which the Atomic Energy Commission now has very little to do. We believe that the planning and funding of civilian nuclear energy activities should now be consolidated with other energy efforts in an agency charged with the mission of insuring that the total energy resources of the nation are effectively utilized. The Atomic Energy Commission would still remain intact, in order to execute the nuclear programs and any related energy research which may be appropriate as part of the overall energy program of the Department of Natural Resources.

Until such time as this new Department comes into being, I will continue to look to the Energy Subcommittee of the Domestic Council for leadership in analyzing and coordinating overall energy policy questions for the executive branch.

Conclusion

The program I have set forth today provides the basic ingredients for a new effort to meet our clean energy needs in the years ahead.

The success of this effort will require the cooperation of the Congress and of the State and local governments. It will also depend on the willingness of industry to meet its responsibilities in serving customers and in making necessary capital investments to meet anticipated growth. Consumers, too, will have a key role to play as they learn to conserve energy and as they come to understand that the cost of environmental protection must, to a major extent, be reflected in consumer prices.

I am confident that the various elements of our society will be able to work together to meet our clean energy needs. And I am confident that we can therefore continue to know the blessings of both a high-energy civilization and a beautiful and healthy environment.

Annex II

ELECTRICAL ENERGY NEEDS AND
ENVIRONMENTAL PROBLEMS IN THE COMMUNITY (E.E.C.)

A note presented to the Colloquium by the Commission of the European
Communities. June 9, 1971

Introduction

Electrical energy needs and environmental problems interest the Commission of the European Communities for several reasons. First, the Community must ensure that its electrical energy requirements are met under the
best possible economic conditions. Second, Article 40 of the Euratom
Treaty requires that the Commission publish periodical illustrative programmes concerning the targets for nuclear energy production and the
various investments involved in implementing them. Third, the Euratom
Treaty also contains a complete chapter on the protection of the general
public and workers against dangers arising from ionizing radiations. The
Commission recently decided to create the necessary administrative infrastructure to study the environment in a more general light and will soon
send the Council the draft of an action programme.

Electrical Energy Requirements

The rapid growth of the population and the accelerated economic expansion
in Europe in recent years have been accompanied by an increase in energy
requirements. Industrial consumption of energy has quadrupled in the last
50 years and is likely to double again in the next 20 years.

Because of simplicity of use, electrical energy is occupying an increasingly
important position among the various energy sources. Its expansion is
anticipated in such areas as steel, the metal industry, air conditioning, town
planning, etc., mainly because of its relatively easy transmission and its
cleanness.

World production of electricity currently exceeds 4 TWh per year, or
more than 1,000 kWh per person. In the Community, the average annual
rate of increase in electrical energy is 7.5%, which represents a doubling of
requirements every ten years. According to the forecasts in the Community's
Second Target Programme (forthcoming) the Community will consume the
following quantities of electrical energy between now and the end of the
century:

1970	559 TWh
1980	1,130
1990	2,080 ⎱ long-term estimates
2000	3,450 ⎰

To meet these requirements, gross production of electrical energy per year should be:

1970	581 TWh
1980	1,210
1990	2,250 ⎱ long-term estimates
2000	3,750 ⎰

Power plants have been gradually supplanting the main indigenous energy source, coal, with oil and natural gas. The resulting heavy dependence on fuels of non-member countries is bound to increase in view of the enormous quantities of electrical energy required in the future and the impossibility of developing the necessary domestic supplies.

According to the Second Target Programme, the respective future functions of the various energy agents capable of ensuring Community electricity production reliably are the following:

Electricity production	Forward estimates			
	1970	1980	1990	2000
hydraulic	118	142	152	157
geothermal	3	3	3	3
thermal (lignite, rec. gas)	92	130	132	120
Privileged production	213	275	287	280
Community coal	131	127	98	42
natural gas	47	150	155	145
imported fuels ⎱ nuclear energy ⎰	190	658	1,710	3,283
Competitive production	368	935	1,963	3,470
Gross production	581	1,210	2,250	3,750

The above table shows the rapid decline in the sources known as "privileged" in Community electricity production to the advantage of competitive sources.

The rationalization programmes for Community coal have resulted in a constant fall in production which is likely to persist for several years. Since the substitution of other fuels for coal is continuing in a number of important sectors, electric power plants and steelworks have become the main outlets for available supplies. Even though support measures have been necessary to ensure the marketing of coal used in power plants, some coal production is still warranted in the Community insofar as it has a favourable influence on Community coking coal production and helps to ensure reliable supplies. Coal Import may increase temporarily because of the rise in the prices of heavy fuel oil. It is probable, however, that no further decisions will be made to construct new exclusively coal-consuming electrical plants in the Community. Under these circumstances, energy production using coal is bound to decrease in time.

The Community's natural gas supply potential has considerably increased in recent years. The geographical concentration of sources certainly represents an appreciable reliability factor for the Community. The Community's natural gas consumption, which amounts to 8 % of total domestic requirements, will therefore certainly increase. Natural gas, however, will above all try to win markets in which its specific advantages make it more economical. In the Netherlands there is a growing tendency to give natural gas an important place in electricity production. A few power plants there are fed exclusively with natural gas and are directly linked to a gas deposit which is shared by the electricity producers. Since coal and natural gas cannot suffice any more than the privileged sources quoted above to produce the additional electricity annually required, their relative contribution will decline.

The imported fuels, largely oil products, together with nuclear energy will meet the future demand for electricity. The price of fuel oil, together with other factors of an industrial nature, has till now been very influential in the development of nuclear energy. Encouraged by the stable and competitive prices prevailing for over ten years, fuel oil has come to dominate the electric power plant market and hold back the rise of nuclear energy. Despite this unfavourable context, nuclear energy has managed to improve its competitive position gradually; partly, perhaps, owing to the rise in price of fossil fuels, but also because of the constant improvement in its own reliability and performance. In the course of time the shrinking costs of nuclear fuel should offset the heavy capital costs inherent in advanced techniques. Though this hard struggle for competitiveness has so far been a major objective, it nevertheless represents only an initial stage. It is not the ultimate goal of nuclear energy to maintain a precarious balance with oil. On the contrary, by aiming at very low prices, it tends to reverse finally and irrevocably the present competitive relationship between conventional and nuclear equipment.

The year 1970 was marked by an energy crisis arising from a combination of a number of factors affecting both supply and demand. The result was a sharp increase in the price of fuel oil plus a threatened shortage. Sooner than anticipated, nuclear energy found itself able to exercise a considerable influence on the future consumption of oil products. It had been acknowledged that until it became competitive, nuclear energy could play only a complementary role, with fuel oil retaining first place in the supply of power plants.

Since nuclear energy at present offers a real economic alternative, i.e., the cost of competing energy sources is high enough for investment decisions no longer to risk being jeopardized every time prices fluctuate, it may be considered that nuclear power plants are capable of developing under other constraints than industrial capacity, grid requirements, financial possibilities, fissile material supplies, etc.

According to the Second Target Programme, the trend in gross annual production of nuclear electricity will be:

1970	15 TWh	
1980	245	
1990	1,090	long-term estimates
2000	2,575	

Electric Power Plants and Pollution

In the past, cost considerations have dictated the production of electrical energy. It will now have to conform increasingly to the rules of environmental protection to enable production to go on developing without detriment to the environment and biosphere.

Conventional and nuclear power plants have different effects on air pollution. Conventional plants discharge a large amount of waste into the atmosphere, whereas nuclear plants are "clean" from this point of view.

For example, a conventional electric power plant of 1,000 MWe operating for 6,500 h/year discharges the following waste:

Tons/year	Coal	Fuel	Natural gas
Sulphur oxides	140,000	50,000	14
Nitrogen oxides	20,000	22,000	12,000
Carbon oxides	500	9	—
Hydrocarbons	200	650	—
Aldehydes	55	120	32
Ash	4,500	725	430

Of this waste, sulphur, carbon and nitrogen oxides and ash are some of the most widespread air pollutants and the ones which, in the opinion of the international organizations, should receive immediate attention. These pollutants should be studied separately and in combination. The problem of their effects is made more complex by the possibilities of mixing, since man is an integrator of pollution and it is the overall effects which must be taken into consideration for the purpose of assessing the harmful effects of air pollutants.

As regards thermal pollution, i.e., the heating effect of the cooling water from power plants, a nuclear plant discharges more calories than a fossil fuel plant. For example, the thermal balance of these two types of power plant is as follows per unit of calorific power produced in the fuel:

	Thermal plant	Nuclear plant
Electric power	0.38	0.31
Stack losses	0.14	—
Miscellaneous losses	0.03	0.04
Losses to river	0.45	0.65
	1.00	1.00

It is obvious that these discharges, for all installed electric power plants make up an enormous loss and that it is necessary to control the choice of sites in accordance with the cooling facilities offered by marine currents, lakes and rivers. It is the opinion of many that the heat discharged should raise the temperature of the zone known as the equilibrium zone (5–10 km downstream of the discharge) by only a few degrees in order to safeguard aquatic biological equilibrium.

Community Action

What steps have the European Communities taken in the past to combat pollution and what is likely to be done in the future? When the various treaties establishing the European Communities were signed, they did not give the Communities special responsibilities in environmental matters because at that time the problem was not as acute as it is now. The European Coal and Steel Community, however, from its inception decided to combat the pollution caused by mines and steelworks and to set aside a percentage of its own resources for this purpose.

The Euratom Treaty, signed at Rome in 1957, contains a complete chapter on the health protection of the general public and workers against the dangers arising from ionizing radiations. The Treaty provides for the establishment of basic standards relating to maximum permissible doses, maximum permissible levels of exposure and contamination, and the basic principles governing the medical supervision of workers. These basic standards were drafted by the Commission with the assistance of national experts and proposed to the Council, which adopted them in 1959. Each Member State then laid down laws and regulations to ensure compliance with these basic standards. Under Article 33 of the Euratom Treaty, the Commission had the right to make any recommendations designed to bring harmony among the provisions applicable in this field in the Member States.

These measures establishing basic standards and their integration into national laws are backed by a series of provisions on the supervision of compliance, and on that of the radioactivity level in the air, water and soil. Thus, according to the Euratom Treaty, each Member State must create the necessary facilities for such monitoring and the Commission has the right of access to these facilities and may examine them with regard to their operation and efficiency. Information on these checks must be communicated regularly by the appropriate authorities to the Commission so that it is kept informed of the radioactivity level to which the population is exposed.

In addition, each Member State is bound to provide the Commission with such general data of plans for the disposal of any type of radioactive waste that could possibly result in radioactive contamination of the water, soil or airspace of another Member State. The Commission, having consulted a group of experts, is bound to express its opinion on these data within six months.

Finally, any Member State on whose territory particularly dangerous experiments are to take place has to adopt additional health and safety measures and must first request the Commission's opinion on them. This opinion must be confirmed when the consequences of such experiments are liable to affect the territories of other Member States. The provisions for checking compliance with the basic standards have been functioning remarkably well for some ten years. However, should a Member State fail to comply with the Commission's recommendations on the level of radio-activity in the air, water and soil, the Commission may adopt a directive enjoining the Member State concerned to take all necessary measures, within a period laid down by the Commission, to prevent the basic standards from being exceeded and to ensure compliance with regulations. If the State in question fails to comply with the Commission's directive within the alloted time period, the Commission, or any Member State concerned, may refer the matter immediately to the Court of Justice.

Although the Treaty establishing the European Economic Community contains no specific provision for protecting the environment this Community began to consider pollution when its programme for the removal of trade barriers was implemented. The General Programme adopted by the Council of Ministers in May 1969, is now being implemented. It contains a *status quo* clause which obligates each state to inform the Commission and the other states of any new measure it intends to adopt in the vast field covered by the Programme and to postpone adoption if necessary, to permit the drafting of Community rules.

The standardized Community regulations do not apply to electric power production, but the Commission is preparing an action programme on the effective protection of the environment and recommended legislation of the utilization of water and air (EEC Treaty Article 100). The action programme in preparation will include proposals concerning the fight against pollution, striving for common criteria, indices and standards, the establishment at Community level of bodies for environmental inspection and management, the implementation of research programmes and various specific schemes aimed at geographical areas or certain industries. This action is to be supplemented at the same time by coordination of the methods of evaluating and monitoring air and water pollution.

Contacts have already been set up with Community institutes to obtain information on the steps currently being taken concerning air and water. Action similar to that conducted under Articles 35 and 36 of the Euratom Treaty could be put into affect without difficulty. Routine monitoring would enable the Commission to be constantly informed of the development of pollution and of the demarcation of certain different zones of pollution, thus initiating the potential establishment of regions of pollution as they are envisaged in other states, notably the United States. Such regions, which would ignore national boundaries would be determined by the results of measurements carried out with standard methods and equipment.

This would then be the first practical attempt to organize the monitoring of air and water quality; the effort at rationalization and standardization necessary to ensure the efficient development of energy production, respect for the environment and balanced conditions of competition might then show concrete results.

INDEX OF AUTHORS AND SUBJECTS

(note) = footnote

252